Coastal Research Library

Volume 33

Series Editor
Charles W. Finkl
Department of Geosciences
Florida Atlantic University
Boca Raton, FL, USA

The aim of this book series is to disseminate information to the coastal research community. The Series covers all aspects of coastal research including but not limited to relevant aspects of geological sciences, biology (incl. ecology and coastal marine ecosystems), geomorphology (physical geography), climate, littoral oceanography, coastal hydraulics, environmental (resource) management, engineering, and remote sensing. Policy, coastal law, and relevant issues such as conflict resolution and risk management would also be covered by the Series. The scope of the Series is broad and with a unique cross-disciplinary nature. The Series would tend to focus on topics that are of current interest and which carry some import as opposed to traditional titles that are esoteric and non-controversial.

Monographs as well as contributed volumes are welcomed.

More information about this series at http://www.springer.com/series/8795

Stuart E. Hamilton

Mangroves and Aquaculture

A Five Decade Remote Sensing Analysis of Ecuador's Estuarine Environments

Springer

Stuart E. Hamilton
Department of Geography and Geoscience
Salisbury University
Salisbury, MD, USA

ISSN 2211-0577 ISSN 2211-0585 (electronic)
Coastal Research Library
ISBN 978-3-030-22239-0 ISBN 978-3-030-22240-6 (eBook)
https://doi.org/10.1007/978-3-030-22240-6

This Springer imprint is published by the registered company Springer Nature Switzerland AG.
The registered company address is: Gewerbestrasse 11, 6330 Cham, Switzerland

While I was in Ecuador conducting research for this book in April 2016, northern Ecuador was struck by a 7.8 magnitude earthquake centered near Cojimies Estuary. I want to dedicate this text to the families of the 676 people who died in the earthquake, the 17,000 that were injured, and the tens of thousands who lost their homes.

Preface

This book has two primary aims:

1. To tell the land use and land cover change story of Ecuador's estuaries from the advent of commercial shrimp aquaculture to the present day
2. To provide a consistent and repeatable remote sensing synthesis methodology that allows for others to conduct such analysis in different regions using different sensors that can be used in both research and instructional settings

This idea for this research began during my first trip to coastal Ecuador in late 2006 and early 2007. During this trip, I traveled up and down the Pacific Ocean coastline between Guayaquil in Guayas Province and San Lorenzo in Esmeraldas. During the long road trips, I saw the magnitude of the landscape changes that had occurred since the advent of commercial aquaculture in this region. I was surprised I did not know about such a massive and recent tropical land cover change. Massive shrimp pond systems dominated the coastal estuaries, while mangrove forests had retreated to a few locations in some isolated locations. Local fisherfolk told me a consistent story of mangrove forest loss and shrimp farm expansion occurring over the prior two or three decades and how now this land cover transition had altered their way of life.

On my return to the USA, I reviewed the academic and broader literature and could only find a limited accounting of what had occurred in this region as it pertained to mangrove loss and shrimp farm expansion. What I did find was a call to arms to tell the story from others interested in this topic. During this time, I had been considering applying to a doctoral program but had yet to find a topic that genuinely motivated me, but upon visiting Ecuador one more time and completing my literature review, I was sure this would be that topic, so I applied to Ph.D. programs with an outline of my dissertation already written. I wanted to tell the land use and land cover change story of Ecuador's coastal estuaries since the arrival of shrimp aquaculture.

I graduated with a Ph.D. about 5 years later with a dissertation entitled, "The Impact of Shrimp Farming on Mangrove Ecosystems and Local Livelihoods Along the Pacific Coast of Ecuador." Although proud of the dissertation, I immediately felt it only told a small portion of the complete story I had been researching. I began to flesh out more of the story in other venues. I was fortunate that others found the research worthwhile and a raft of publications followed including pieces in *Nature Climate Change*, *Annals of the Association of American Geographers*, *PLOS One*, *Global Ecology and Biogeography*, *Bosque*, and *Proceedings of the National Academy of Sciences of the United States of America* and a chapter in the Coastal Research Library, to which this book belongs. All these manuscripts told interesting, but isolated, portions of the bigger story. This initial mangrove and shrimp farm research agenda additionally provided me both fellowship and further research opportunities in Ecuador to continue researching the topic. As my mangrove research expanded over time, the desire to revisit the entire coastline of Ecuador and attempt to write the definitive text grew stronger. Despite the journal articles, presentations, and book chapter publications, the feeling that the full story was not being told persisted within me, and I began to realize that a comprehensive monograph was likely the only means to get the full story out to other academics and hopefully both the public and policy-makers.

From 2006 to present, I continued expanding on my research into mangrove loss and shrimp farm expansion and visited Ecuador every year for almost a decade, including living on the coast in both 2005 and 2016. Among other activities, I participated on a working shrimp farm, hiked through remote mangrove forests, and visited a shrimp hatchery. While in Ecuador, I visited the remaining mangrove forests that I had been unable to reach prior. From 2006 to 2016, I conducted interviews with fishing syndicates, held meetings with local researchers, conducted ethnographic research, and held PPGIS workshops all along the coast. Sarah Collins took the lead on the livelihood research while I kept expanded the land cover and land use theme of the research. Ramon Cedeno Loor took me under his wing and arranged for visits to ever more remote mangrove forests and meetings with more stakeholders. With the help of many students, I processed more land cover data from ever more resolute Earth observing systems and incorporated these data into my ever-growing spatial databases. By early 2016, the book was ready to be written as I finished my fellowship in Bahia de Caráquez, Ecuador.

From 2016 to 2018, the text was written and revised, and Springer accepted the book as part of their Coastal Research Library series. The book attempts to give as complete as possible account of the land cover and land use change that occurred in Ecuador's estuaries from the advent of commercial aquaculture to present. The book covers each estuary in Ecuador to allow for local, regional, provincial, and national results to be derived. In addition to telling the Ecuador story, the book presents a methodology of remotely sensed data synthesis that can be applied to other regions and other land cover types globally.

The first chapter of the book is dedicated to mangrove forests. It introduces mangrove forests and examines mangrove biology including mangrove adaptive mechanisms that allow them to exist in saline and anaerobic environments, their evolution, their current distribution, their paleodistribution including the mangrove anomaly, their historic and current coverage area estimates, their traditional uses, their species composition, their support of fisheries, their management regime, and their role in climate change. The chapter examines each of these topics at the global scale, at the Ecuadorian national scale, and when possible at the local scale. The chapter is written for coastal researchers who not are either biologists or botanists but desire to gain a foothold in mangrove research and understand the role of mangrove forests.

The second chapter of the book provides a review of shrimp aquaculture. It provides a background on shrimp farming and wider aquacultural practices. The chapter is again written for none aquaculture experts to gain the required knowledge of shrimp aquaculture to conduct such analysis as this in other regions. This chapter covers the biology of shrimp including the complex lifecycle of *Penaeus vannamei* in both the wild environment and farmed systems. Additionally, it covers the operation and management of shrimp farms, the legal framework of aquaculture in Ecuador, and the importance of Ecuador's estuaries and mangrove forests to both farmed and wild shrimps. Again, the chapter transitions from global to Ecuador to local in scale.

The third chapter of the book delineates the study sites and subdivides the coastline of Ecuador into logical estuarine units for analysis. It is dedicated to the local situation. The chapter provides information on each of the study sites including the climatic conditions of each of the estuaries, the environmental regime present in each estuary, the protected status of each estuary, the biodiversity associated with each estuary, and the socioeconomic condition of the residents who reside near the estuary. It not only attempts to provide the reader with the statistics, facts, and boundaries for each estuary but also attempts to additionally provide information on the milieu of the residents of the estuaries including information on the lifestyle, livelihoods, and culture of each estuarine community.

The fourth chapter is dedicated to the methodology with a significant portion of the chapter describing remote sensing of the Ecuadorian estuaries from pre-1970 to present. It presents a consistent and widely applicable remote sensing methodology that allow for repeating this study in Ecuador and more importantly to duplicate this study in other regions. It reviews the Earth observing instruments used, the feature delineation approaches taken, the GIS tools applied in the method of change detection, and how the differing Earth observing systems and other data were combined into a singular spatiotemporal analysis. The data required to repeat this approach is provided in the accompanying Dataverse.

The fifth chapter presents the results of the study. The results are presented at the estuarine level and then summarized to the national level. The first set of results presented document the changes in areal extent of mangrove forests and shrimp farms across all study sites. The second set of results examines the displacement of one land cover type by another. The results also present the spatiotemporal land cover changes and the sub-estuary and estuary scale. Finally, the estuary results examine the patterns revealed in the remote sensing analysis that validate the oral histories provided by local stakeholders. The data required to repeat this approach is provided in the accompanying Dataverse.

Finally, the sixth chapter of the book is a brief discussion that identifies future research needs and identifies data gaps, presents some potential policy ideas for consideration, and discusses the broader implications of this study at multiple scales. The discussion also describes the pros and cons of the methodological framework presented and suggests additional regions that may benefit from a similar approach.

The accompanying Dataverse provides all data in GIS and tabular formats. This includes all input data, all output data, and other data required for data processing. The URL is https://dataverse.harvard.edu/dataverse/MFACRL.

Salisbury, MD, USA Stuart E. Hamilton

Acknowledgments

I want to thank Dr. Klaus Meyer-Arendt for initially introducing me the coastal region of Ecuador and providing funding for our initial trip up and down the coast. I also wish to thank Donald Hagan and Tia Hagan for introducing me to the mangrove sites in the Rio Chone Estuary that first garnered my interest in this topic.

I want to acknowledge the Nature Conservancy; Secretary of Higher Education, Science, Technology, and Innovation (Senescyt); the University of Southern Mississippi; the University of West Florida; and Salisbury University for funding portions of this research.

Thank you to Clay Harris, Justin Fitt, Christopher Kaempf, Sarah Collins, Marco Millones, Victor Collins, Shauna Kiernan, Mike Mitchell, Ramon Cedeno Loor, Jose Eduardo Ganchozo, Sean Conner, Nathan McKinney, Don Hagan, Erica Moldenhauer, Jose, Leonardo Caeua, Ronald Zembrano, the field staff of Global Student Embassy, the staff of the local MAE office in San Vicente, and countless others for providing their labor free of charge while in the field.

Thank you to Clare Stankwitz, John Lovette, Sarah Byrd, and all the Salisbury University geography students for their help with digitizing and extracting land use and land cover data for the estuaries.

I would like to thank the fisherman of the Isla Corazon community for allowing me to interview the cooperative members and for providing numerous trips into the mangrove forest.

I would like to thank Clare Stankwitz for her help on compiling the shrimp farm financing data used in Chap. 2.

I would like to thank John Talbot for automating of tasks within this process.

I would like to thank Liam Hamilton for summarizing global mangrove numbers and cross-checking math operators.

I would like to thank Paige Roberts for her editorial assistance and reviewing Chap. 2 as well assisting with the remote sensing systems timeline graph.

I would like to thank Noah Krach, Gustavo Castellanos, and Paige Roberts for reviewing each of the chapters of this book and providing throughout insights to improve the text.

Without funding and data availability decisions made by the NASA, the USGS, the Government of Japan, and other agencies, then, research of this type would become unaffordable.

Thank you to Sarita Collins for the translation services, assistance in the field from her and her extended family, helping develop the ideas for this book, and always being available at short notice with a team of workers to assist in the field. She not only assisted but applied her local knowledge to help me gain access to many local communities as well as fulfill my other important roles while in the field. She is the only other individual aside from Ramon Cedeno Loor and I involved from start to end in this research process, and without her, this research would not have been possible.

I would like to thank Noah Krach for cleaning the data and helping process the estuarine data. Noah is a coauthor on the published data on the Dataverse. Without his effort, Chap. 4 would have extended on indefinitely.

Finally, I would like to thank my good friend Ramon Cedeno Loor for the 13 years of support in Ecuador and beyond. He introduced me to the local fisherman, took me to the major shrimp farms, and was always happy to set up meetings with public officials. He provided many of the insights that are fleshed out in this text. He traveled with me every trip and worked every day to achieve the environmental goals of the research. He ensured my safety in dangerous areas and worked tirelessly everyday to help me achieve the goals of this manuscript. His dedication to the ecological health of the region is second to none, and without him, this manuscript would be no more than an unfulfilled idea.

Contents

Acronyms

AC	Agulhas Current
ACEP	Atlantic, Caribbean, and Eastern Pacific
ASTER	Advanced Spaceborne Thermal Emission and Reflection Radiometer
BC	Brazilian Current
BeC	Benguela Current
BMN	Baculoviral Midgut Gland Necrosis
BOD	Biological Oxygen Demand
CE	Chone Estuary
CI	Confidence Interval
CJ	Cojimies
CLIRSEN	Centro de Levantamientos Integrados de Recursos Naturales por Sensores Remotos
CM	Cayapas-Mataje
DAC	Development Assistance Committee
DBH	Diameter at Breast Height
DEM	Digital Elevation Model
EAC	East Australian Current
EOS	Earth Observing System
ESA	European Space Agency
ETM+	Landsat Enhanced Thematic Mapper Plus
FAO	Food and Agriculture Organization of the United Nations
FARC	Revolutionary Armed Forces of Colombia
FCR	Feed Conversion Ratio
FDI	Foreign Direct Investment
FUNDECOL	Fundación para la Defensa Ecológica
GDP	Gross Domestic Product
GFC	Global Forest Cover
GG	Gulf of Guayaquil
GIS	Geographic Information System
GLCF	Global Land Cover Facility

GS	Gulf Stream
HAB	Harmful Algal Bloom
HC	Peruvian/Humboldt Current
IFC	International Finance Corporation
IGM	Instituto Geografico Militar
IHHNV	Infectious Hypodermal and Hematopoietic Necrosis
IMF	International Monetary Fund
ISME	International Society for Mangrove Ecosystems
ITTO	International Tropical Timber Organization
IUCN	International Union for Conservation of Nature
IWP	Indo-West-Pacific
KC	Kuroshio Current
LULC	Land Use and Land Cover
MAE	Ministerio del Ambiente del Ecuador
MFW	Mangrove Forests of the World
MHHW	Mean Higher High Water
MLC	Maximum Likelihood Classification
MS	Muisne
MSS	Multispectral Scanner 1
NASA	National Aeronautics and Space Administration
NDVI	Normalized Difference Vegetation Index
NGO	Nongovernmental Organization
NOAA	National Oceanographic and Atmospheric Administration
OECD	Organisation for Economic Co-operation and Development
OLI	Operational Line Imager
PBDE	Polybrominated Diphenyl Ethers
PCB	Polychlorinated Biphenyls
PCDD	Polychlorinated Dibenzodioxins
PCDF	Polychlorinated Dibenzofurans
PMRC	Coastal Resources Management Program
PPGIS	Public Participation Geographic Information Systems
PSAD	Provisional South American Datum
RBV	Return Beam Vidicon
REMCAN	Reserva Ecológica Manglares Cayapas Mataje
REye	RapidEye
RMSE	Root-Mean-Square Error
SLC	Scan Line Corrector
SNAP	Sistema Nacional de Áreas Protegidas del Ecuador
SPOT	Satellite Pour l'Observation de la Terre
SWIR	Short-Wave Infrared
TIR	Thermal Infrared
TM	Landsat Thematic Mapper
TOA	Top of Atmosphere
TSV	Taura Syndrome Virus
UAV	Unmanned Aerial Vehicle

UN	United Nations
USAID	United States Agency for International Development
USGS	United States Geological Survey
UTM	Universal Transverse Mercator
VHR	Very High-Resolution Satellite-Based Optical Imagery
VNIR	Visible and Near-Infrared
WDPA	World Database on Protected Areas
WGS	World Geodetic System
WRS	World Reference System
WSSV	White Spot Syndrome Virus
ZEM	Special Area Management Plan

Chapter 1
Botany of Mangroves

To the land animal or plant, the sea is a hostile environment. High salinity, wave action, and fluctuating water levels present problems that are rarely experienced in terrestrial or freshwater habitats.

(Hogarth 2007 P. 1)

Abstract Despite such harsh environmental conditions, mangroves have colonized the nearshore environment of much of the global tropical coastline. In doing so, mangrove forests have converted what is otherwise often barren coastlines into one of the most productive ecosystems on the planet. Measured regarding human livelihoods, carbon processing, fisheries support, erosion control, or biodiversity; mangrove forests are one of the most productive ecosystems on the planet, yet they mangrove forests are historically one of the most underappreciated and misunderstood land cover types. This chapter examines the botany of mangroves. Section 1.1 describes the biology of mangroves with a focus on those present in Ecuador. Section 1.2 examines the dominant Ecuadorian mangrove that is *Rhizophora mangle* which is currently undergoing a taxonomic reclassification to *Rhizophora samoensis*. Section 1.3 delineates the global and Ecuadorian distribution of mangrove species. Section 1.4 examines current and past estimates of mangrove forest area for both global mangrove forests and Ecuadorian mangrove forests. Section 1.5 examines the role of mangrove forests in supporting the wider ecosystem that they inhabit. Section 1.6 focusses upon the role of mangrove forests in supporting wild fish fisheries and Sect 1.7 on the manner in which mangrove forests provided important livelihood options and other goods and services to the people that reside nearby. Section 1.8 elucidates on the role of mangrove forests in mitigating global climate change and recent advances in this critical arena.

Electronic Supplementary Material The online version of this chapter (https://doi.org/10.1007/978-3-030-22240-6_1) contains supplementary material, which is available to authorized users.

S. E. Hamilton, *Mangroves and Aquaculture*, Coastal Research Library 33,
https://doi.org/10.1007/978-3-030-22240-6_1

Keywords Mangrove forests · Mangroves · *Rhizophora mangle* · *Rhizophora samoensis* · Blue carbon · Mangrove economy · Mangrove forest cover · Mangrove anomaly

1.1 Mangrove Forest Biology

Mangrove, mangrove forest, mangrove forest trees, and mangal are used interchangeably in the geographic literature to describe the tree, plant, and shrub vegetation that exists in and around the tropics. Mangrove forests are most often located at the ocean and land interface, within secluded swamps, and at the mouths of rivers in estuarine environments. The term mangal usually represents the entire flora present in a mangrove forest swamp, and the term mangrove forest refers to those species within mangal that are taxonomically now classified as mangrove forest (Macnae 1968). Although somewhat redundant, this appears to be the most consistent use of this terminology in the botanical literature. Mangrove itself is likely a compound of the Portuguese word *mangue,* from the now extinct indigenous Taino language of the Bahamas and the Greater Antilles, and of the English term *grove.* An alternate but less accepted origin is that mangrove is an adjustment of the Malayan *manage-manggi* term which is still used in eastern Indonesia, although no longer in Malaysia, to describe the local *Avicennia* species (Bakhuizen van den Brink 1921).

There is now general agreement on the species that are considered mangroves, although new species and hybrids are still being classified. The rationale behind the grouping differs from the typical practice of basing biological taxa on common ancestry. The ambiguous mangrove forest grouping is likely because mangroves have not evolved from a single or even a handful of species, but have converged into similarity due to adaptations from inhabiting similar environments. Hence, mangrove forest taxonomy utilizes the environmental conditions of the flora habitat to delineate what species constitute mangrove forests. As such, species labeled as mangrove forest do not necessarily belong to the same taxa; for example, the seven species of mangrove forest within Ecuador belong to five different families and only two species of Ecuadorian mangrove forest belong to the same genus. Classification of mangrove species is thus in many ways a circular process, since the term mangrove forest is generally applied to plant species with specific adaptations that inhabit tropical tidal swamps, yet tidal swamps close to the tropics are often typically defined by the fact that they have mangrove forest present within them (Tomlinson 1986; Hogarth 1999, 2007; Hogarth 2015). Additionally, diverse flora often exists on the terrestrial fringe of intertidal swamps but are omitted from the general classification of mangrove forest because these species have a more dominant presence in terrestrial environments and taxa.

Several common characteristics generally exist in many species referred to as mangrove: (i) A physical adaption to the anaerobic and waterlogged environment; (ii) an adaptive mechanism for living in highly saline water and highly saline soils; (iii) taxonomic isolation from terrestrial relatives, (iv) geographic fidelity, (v) and the presence of a vivipary/crypto-vivipary embryonic structure. We examine each of these in more detail below.

(i) The primary mangrove forest adaptation that allows them to flourish in both waterlogged and anaerobic soils is adaptive root structures. At least four different mangrove forest root adaptations have been noted, and these are termed as pneumatophores, prop or stilt roots, kneed roots, and plank roots (Kelvin et al. 2001). Many species of mangroves roots allow for efficient respiration through hundreds of tiny pores known as lenticels that remain above the water level or close to the water level during high tides (Yáñez-Espinosa and Flores 2011). For example, the *Rhizophora mangle* in Ecuador has stilt roots that are often many meters above the waterlogged and saline soils below (Fig. 1.1).

(ii) Mangroves' adaptive mechanisms for existing in highly saline environments vary across species but can broadly be classified into three overlapping categories. First, many mangrove species roots are effective at filtering out much of the salt water before it enters the plant. Such root processing is likely via a process of ultrafiltration that blocks the uptake of almost all of the salt ions present (Kelvin et al. 2001). Second, many mangroves store excess salt in old bark or their leaves; they then litter or shed bark regularly to rid themselves of the accumulated excess salt. Third, many mangroves have evolved the ability to secrete salt through glands in the leaves. Finally, many mangroves are adept

Fig. 1.1 Rhizophora mangle Stilt Roots in Chone Estuary, Manabi, Ecuador
These *Rhizophora mangle* stilt roots extend about 5 m above the highly saline and waterlogged soils. These are atypically large examples and are found in northern Ecuador. The Mean Higher High Water (MHHW) approximately corresponds to the highest prop root in the picture

at storing high levels of fresh water in their thick leaves and limiting evaporation (often due to a waxy leaf coating), thereby retaining fresh water for more prolonged periods and helping balance the level of salt present.

(iii) If a shrub or tree species is present in a saltwater environment and also in a terrestrial environment, but it is more dominant and numerous in the terrestrial environment, it is usually not classified as a mangrove forest tree but often termed a mangrove forest associate. The mangrove forest grouping generally appears to require taxonomic isolation from terrestrial relatives or at least dominance in the aquatic environment as opposed to the terrestrial environment, to be grouped into the "true mangrove" class (Tomlinson 1986). Another proposed method of separation is that mangroves are true halophytes[1] whereas mangrove forest associates are generally glycophytes[2] (Wang et al. 2011). Regardless of the method of division, species classified as mangrove forest do not dominate terrestrial ecosystems and Guevara and Granda (2009) record Ecuador as having six true mangroves and an additional six mangrove forest associates. The Food and Agricultural Organization of the United Nations (FAO) classifies Ecuador as having seven mangrove species (UN FAO 2004).

(iv) Geographic fidelity refers to mangroves' tendency to exist in dense stands composed of a single mangrove species with a geographically clear delineation when transitioning from one region to another. This sorting of species is usually attributed to salinity tolerance levels, although this is likely an oversimplification of the complex processes that result in the homogeneity of species within regions of a single swamp. Other proposals for mangrove forest sorting have suggested that the size, weight, or buoyancy of propagules may play a factor. Within Ecuador, between 80% and 90% of all mangrove forest stands are *Rhizophora mangle* (Hamilton and Lovette 2015; Madsen et al. 2001; Arriaga et al. 1999) with the other 10% consisting mostly of *Avicennia germinans* and *Laguncularia racemosa*.

(v) Many mangroves exhibit vivipary; this is the condition whereby the embryo grows first to break through the seed coat, then out of the fruit wall, while still attached to the parent plant. Vivipary is common in *Bruguiera*, *Ceriops*, *Kandelia*, and *Rhizophora* species (Kelvin et al. 2001). Cryptovivipary refers to the condition in which an embryo grows to break through the seed coat but not the fruit wall before it splits open. Cryptovivipary occurs in *Aegiceras*, *Avicennia*, and *Nypa* species (Kelvin et al. 2001). Although not unique to mangroves, the exact explanation for why so many mangrove species exhibit a form of vivipary remains unknown.

Mangrove forest pollination appears similar to the majority of other plants with limited specific adaptations present. Across differing species, pollination occurs by wind, bees, bats, moths, birds, butterflies, and a host of small insects (Tomlinson 1986). More interesting than pollination is the role of propagules and the process of

[1] A halophyte is a plant adapted to growing in saline conditions, as in a salt marsh.

[2] Any halophyte that can only tolerate relatively low concentrations of salt.

vivipary in mangrove forest reproduction. Across all the academic literature, the reproduction and dispersal method for many mangroves appears to be via dispersal of the propagules dropped into the water. The propagule, or seedling, develops while still attached to the parent, often for many months, at some point, it drops from the parent and floats away, usually for an extended period and often over a long distance. The propagule may or may not eventually root in a suitable substrate and then develop.

The vivipary (*Rhizophora, Bruguiera, Ceriops, Kandelia*) or cryptovivipary (*Avicennia, Aegiceras, Nyoa*) methods differ slightly in how the propagule develops on the parent but then follow much the same process of reproduction and dispersal (Hogarth 2015; Kelvin et al. 2001). Within Ecuador, replanting of the dominant *Rhizophora mangle* often merely involves harvesting large propagules from one location and inserting them into the soil at the replanting location. It is hypothesized that propagules can remain viable while floating in the water for months or even years after they drop from the parent. Other mangroves outside of the species noted above disperse following a conventional seeding process.

Due to their non-traditional taxonomic definition, the number of identified mangrove species varies in the literature. The Tomlinson (1986) mangrove classification system allows for 54 mangrove species in 20 genera and 16 families, whereas other classifications allow for as many as 70 (Spalding et al. 1997) and 80 (Blaber 2007) distinct mangrove species. Table 1.1 identifies the six species of mangrove present in Ecuador (Guevara and Granda 2009). The FAO defines the fern *Acrostichum aureum* as a mangrove species that is present in Ecuador (UN FAO 2004), but this manuscript follows the local tradition in Ecuador and counts this species as mangrove forest associated fauna (Guevara and Granda 2009). Ecuador has four fewer

Table 1.1 Field verified Ecuadorian mangrove forest taxa

Species name	Common name	Family
Avicennia germinans	Black Mangrove, Aili, Calumate, Mangle negro, Mangle salado, Palo de sal	Verbenaceae
Conocarpus erectus	Buttonwood, Button Mangrove, Sea Mulberry Botoncillo, Button mangrove, Buttonwood, Mangle botoncillo, Mangle mariquita, Mangle torcido, Zaragoza, Zaragoza	Comretaceae
Pelliciera rhizophorae	Tea Mangrove, Palo de sal, Pie de Santo	Theaceae
Rhizophora harrisonii	Mangle caballero	Rhizophoraceae
Rhizophora mangle/ *Rhizophora samoensis*[a]	Red Mangrove, American Mangrove, Mangrove, Aili ginnid, Ailikinnut, Mangle, Mangle Colorado, Mangle gateador, Mangle rojo, Mangle salado	Rhizophoraceae
Laguncularia racemosa	White mangrove	Combretaceae

Conocarpus erectus is often described as not a true mangrove species (IFAS 2009). Data adpated from Guevara and Granda (2009)

[a]There is currently confusion between these two species along the Pacific Coast of the Americas

species than western coastal Colombia and five fewer than Panama's Pacific coast. Peru has five mangrove species, all of which exist only near the Ecuadorian border.

The lack of mangrove species diversity in Ecuador is a result of two interacting conditions. First, though it is on the equator, Ecuador is close to the maximum southern extent of mangrove forest along the Pacific coast of the Americas. Mangrove forests rarely exist where the sea surface temperature is colder than 20 °C, and this isotherm is located at approximately S3.75° off the western coast of South America due to the cold Peruvian Current or Humboldt Current. By contrast, the southern extent of mangroves on the Atlantic side of South America is S33°, almost ten times farther south than on the Pacific coastline. Second, the diversity of mangrove species in Ecuador is limited by geographic isolation from the species-rich area of the Indo-West-Pacific region (IWP). The IWP has between 47 and 51 mangrove species, as opposed to only seven in Ecuador. All Ecuadorian mangrove species exist in other parts of the Atlantic, Carribean, and Eastern Pacific (ACEP), although this is in question (Ellison et al. 2015; Ellison and Duke 2015).

The most widely accepted mangrove forest origination hypothesis is a vicariance hypothesis that states that mangroves separated from terrestrial vegetation in the Late Cretacious period in and around the Tethys Ocean (Ellison et al. 1999), and regional differentiation occurred *in-situ* post continental drift (Ellison et al. 1999; Hamilton et al. 2017). Such a hypothesis ties well with the mangrove forest fossil record. For example, the fossil record suggests both *Nypa* (He et al. 2015) and *Acrostichum* (Zhang et al. 2016) diverged from their terrestrial relatives during the Late Cretaceous. Additionally, both *Pelliceria spp.* and the now extinct *Brevitricolpites* have been dated to this period (Plaziat et al. 2001).

After the Late Cretacious, the vicariance hypothesis points to a trifecta of independent regional diversification in Southeast Asia, the Indian Ocean, and the ACEP (Ellison et al. 1999). Such a hypothesis is supported by the limited mangrove species found in Ecuador which are all also found through the ACEP. As the mangrove species of Ecuador are consistent with mangroves across the ACEP as opposed to the IWP, it can be inferred that the *Rhizophora mangle*-dominated coastline of Ecuador, and indeed of the entire Eastern Pacific, most likely developed before the formation of the Isthmus of Panama. The formation of this isthmus is generally agreed to have occurred approximately 2.8 Ma (O'Dea et al. 2016). Any changes in the mangrove forest structure of Ecuador and the Eastern Pacific occurring in the last 2.8 million years is, therefore, most likely *in-situ* diversification.

1.2 Rhizophora Mangle & Rhizophora Samoensis

Due to its dominance in Ecuadorian estuaries, it is worth expanding on *Rhizophora mangle* (Fig. 1.2). Although six or seven mangrove species are present in Ecuador (Table 1.1), *Rhizophora mangle* is the dominant species (Hamilton et al. 2017;

Fig. 1.2 Rhizophora mangle
Panel *1* represents the leaf (b), flower (a, b), fruit (c) and the propagule (d) of the *Rhizophora mangle* (Zorn 1796). Panel *2* represents the tree at an extreme latitude; Panel *3* represents the extensive prop root system of two mature trees on or near the equator, Panel *4* represents the exterior of a *Rhizophora mangle* stand on or near the equator. Panels' *3* and *4* photographs are taken by the author in Ecuador, Panel *1* is constructed by Johannes Zorn and is copyright expired, Panel *2* is photographed by Hans Hillewaert in Central Florida and released under a CC BY-SA 3.0 license

Arriaga et al. 1999; Spalding et al. 1997; Spalding et al. 2010; Madsen et al. 2001), with estimates that *Rhizophora mangle* constitutes as high as 90% of all Ecuadorian mangrove forests (Hamilton and Lovette 2015). Hamilton et al. (2017) recently sampled random plots across Ecuador's northern estuaries, and 80% of these random plots were found to be *Rhizophora mangle*. In the largest estuary in northern Ecuador, *Rhizophora mangle* constituted 98 percent of all the random sites.

As of 2007, *Rhizophora mangle* is listed by the International Union for Conservation of Nature (IUCN) as a species of least concern despite its decreasing status (Ellison et al. 2015). Interestingly, the species is not listed as present in the Eastern Pacific or Ecuador by the IUCN Red List because the Eastern Pacific Red Mangrove has recently been reclassified as *Rhizophora samoensis* and now differs from the historical definition of a singular Red Mangrove, *Rhizophora mangle*, that is present across the entire ACEP (Ellison and Duke 2015). In this manuscript, we use *Rhizophora mangle* L., as the taxon name for historical consistency but it should be noted that shortly the taxon name for all Ecuadorian Red Mangroves will likely be officially reclassified as *Rhizophora samoensis* (Hochr.) *Salvoza*. At this time, *Rhizophora mangle* remains in dominant usage, and pre-2015 *R. samoensis* appears to have never been used to refer to the Red Mangroves of Ecuador aside from a single reference by Twilley et al. (2001). Despite this, it is worth a brief discussion on *R. samoensis* as this is likely the correct taxon.

R. samoensis shares the same class, order, and family as *Rhizophora mangle* and is listed as occuring along the mangrove forest range of the Pacific Coastline of the Americas. It is currently listed as near threatened in the RED LIST v3.1 (IUCN 2010). It also exists in much of Australasia. It is generally regarded as the tallest of all mangrove species, growing up to 60 m in height in Ecuador (Ellison and Duke 2015). Potential threats are listed as urban development, non-timber crops (likely oil palm), aquaculture, wood harvesting (likely for charcoal), and climate change (Ellison and Duke 2015).

The primary method to categorize the distinction between *R. samoensis* and *Rhizophora mangle* is a nuanced difference in the pistil. The portion of the pistil between the ovary and the stigma, know as the style, is typically greater than 2.1 mm in length for *R. samoensis* and less than 2.1 mm for *Rhizophora mangle* (Duke 2018). Additional differences are bracteoles inconspicuous (< 1 mm long) vs. bracteoles conspicuous (1 mm or longer), flower buds rounded to the apex vs. flower buds constricted towards the apex, and *samoensis*-like vs. *mangle*-like (Duke 2018). Finally, field observation by the author of *Rhizophora mangle* along the Atlantic coastline of the USA, Belize, Mexico, Brazil, Haiti, and Dominican Republic compared to field observation of what is now classified as *R. samoensis* along the Pacific Coastline of Colombia and Ecuador, appears to indicate that the extreme heights greater than 40 m are only achieved by *R. samoensis* in the Americas.

1.3 Mangrove Species Distribution

Mangrove forests exist across tropical coastlines where the sea-surface temperature remains consistently above 20 °C, and the local conditions allow (Fig. 1.3). The extreme northern longterm mangrove forest appears to be in Bermuda (Hamilton and Casey 2016) and is likely made possible by the warm waters of the Gulf Stream (Fig. 1.3). No other mangrove forest exist at such extreme northern latitudes. The extreme southern longterm stand appears to be in and around southern Victoria, Australia (Hamilton and Casey 2016) and is likely made possible by the warm waters of the East Australian Current (Fig. 1.3). Mangrove forests exist in similar southern latitudes in New Zealand as those in Australia. No mangrove forests exist in Europe or Antartica (Fig. 1.3).

Despite the latitudinal extremes noted above, the majority of mangrove forests exist within the tropics of the Americas, Africa, Asia, and Oceania (Fig. 1.3). The primary two exceptions to this are the western coast of South America and the southwestern coast of Africa, where areas clearly within the tropics are lacking mangrove forest. The first and most notable exception zone is along the western coast of South America where mangrove forests extend only approximately 300 km south of the equator. The Tropic of Capricorn is over 2000 km farther south of the farthest mangrove forests along this coastline. This limited southern latitudinal range along the west coast of South America is due to the frigid waters of the Peruvian/Humboldt current (Fig. 1.4, HC) that extends from the southern polar region toward the Ecuadorian border with Peru and has water temperatures too cold for mangrove forest establishment. The second exception is not as extreme as the limitation on the western coast of South America but is still notable in Fig. 1.3. The cold water Benguela current limits the southern expansion of mangrove forests towards the Tropic of Capricorn along the Southwestern coast of Africa (Fig. 1.4, BeC).

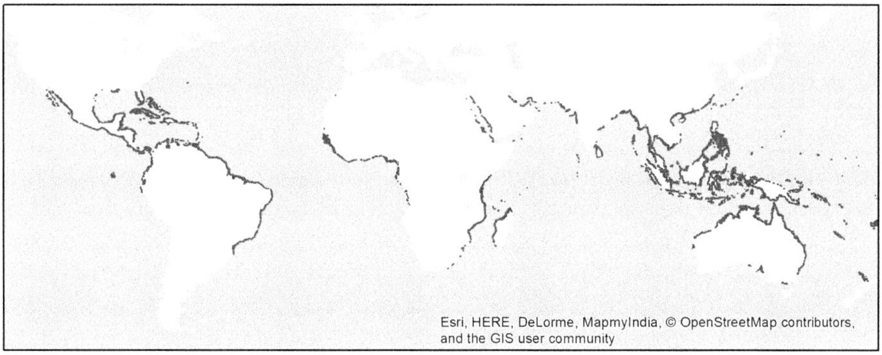

Fig. 1.3 Global distribution of mangrove, circa 2000
Map created by the author with data from Giri et al. (2011). These GIS data are available for download at the Socioeconomic Data and Applications Center (SEDAC) using the URL http://sedac.ciesin.columbia.edu/data/set/lulc-global-mangrove-forests-distribution-2000

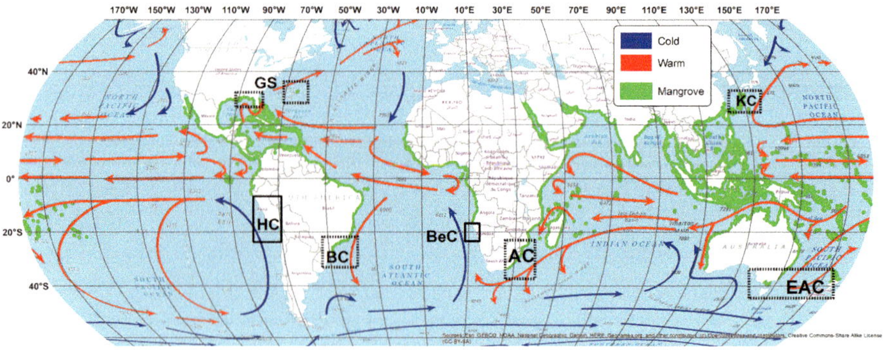

Fig. 1.4 Global Ocean currents and mangrove forest distribution
Map created by the author with data from NOAA and the US Army. GS is the Gulf Stream. HC is the Humboldt Current or Peruvian Current. BC is the Brazilian Current. BeC is the Benguela Current. AC is the Agulhas Current with a contribution from the Southern Equatorial Current. KC is the Kuroshio Current. EAC is the East-Australian Current.

The GIS data for this map are available for download from the accompanying dataverse at https://doi.org/10.7910/DVN/TKGO2Z, and a tutorial on the map creation process is available on YouTube at https://youtu.be/nN8sl7RuXXs

Cold ocean currents can limit mangrove forests extent within the tropics, and therefore conversely, warm ocean currents can expand the latitudinal range of mangrove forests beyond the tropics. In North America, the Gulf Stream (Fig. 1.4, GS) supports mangrove forests as far north as the Bahamas and the panhandle of Florida, and even into Louisiana. While in South America, the Brazilian current allows mangrove forests south of S33° along the eastern coastline of South America (Fig. 1.4, BC). Parts of South Africa far outside the tropics additionally have mangrove forests; again, these are supported by the warm waters of the Agulhas Current and the Southern Equatorial Current. In Asia, mangrove forests can be found as far north as Japan and are supported by the warm Kuroshio Current (Fig. 1.4, KC), and in Australia and New Zealand, the East-Australian Current provides the heat for mangrove forests in Southern Victoria, Australia, and in sheltered bays on New Zealand's north island.

Again, the 20 °C isotherm appears critical to global mangrove forest presence, although the exact reasons for this strict temperature control are not clear. It is hypothesized that the metabolic costs of tolerating salt water, combined with adjusting to both anaerobic and waterlogged soils, while functioning as a tree (as opposed to seagrass for example) is not metabolically practical outside of the tropics where high temperatures and long days increase the potential for photosynthesis (Hogarth 2015). It should be noted though that the exact reasons for the strict temperature control on mangrove forest extent remain unelucidated.

Figure 1.5 depicts the global richness of mangrove species according to the IUCN. The author has subdivided the tropics and near tropical regions of the globe into 5° by 5° grid segments, commonly called a fishnet. The fishnet is then overlaid

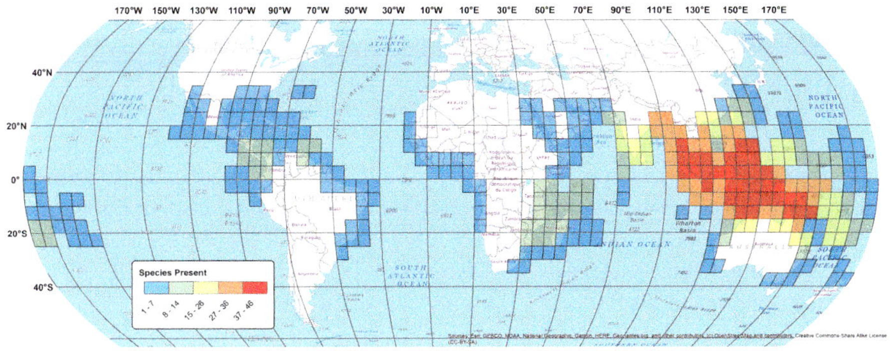

Fig. 1.5 Global distribution of mangrove species
Map created by the author from data provided by the IUCN. Overestimation of species appears present in and around northern South American and southern Central America (Blanco 2018)
These mangrove species distribution GIS data, as well as the compiled mangrove fishnet counts, are available for download at https://doi.org/10.7910/DVN/FK8JZK

on the globe, and the number of different mangrove species present in each cell is aggregated. The resultant map is the number of mangrove species present in each grid segment. In many ways, the construction of the fishnet is arbitrary, but in this use case it starts at S40° and ends at N40°. The start and end parallels were selected to capture the full latitudinal mangrove forest extent, to align with the equator, and to nest within 10° segments, 20°, and so forth. On the longitudinal axis, the grid starts at W180° and ends at E180°, again to capture the full known mangrove forest extent and nest within the significant meridians as well as aligning with the prime meridian.

Moving from east to west across Fig. 1.5, starting along the Pacific Coast of the Americas, the majority of the ACEP region is generally limited to seven or fewer species with a few locations around the Caribbean and West Central Africa having more than seven mangrove species but never more than 14 mangrove species. Continuing east into the East African portion of the Indian Ocean around Mozambique and the island nation of Madagascar, the ACEP pattern inverts with most locations having between 8 and 14 mangrove species but with the latitudinal extremes still limited to seven or fewer mangrove species. Continuing to move from west to east toward the Indian subcontinent, the species richness increases to 15–26 mangrove species. Again, moving farther east around the Bay of Bengal toward Southeast Asia, the mangrove species richness increases to the 27–36 range. Finally, moving into the core of the IWP, the mangrove species richness increases once again to a maximum of 46 species in one location with many 5° segments having 37 or more mangrove species present. The center of species diversity includes many hundreds of the extreme east Indonesian islands and the Indonesian portion of New Guinea; this 5° grid cell has 46 distinct mangrove species (Fig. 1.5). Numerous cells on the latitudinal extremes have only one mangrove species; a cluster of cells in the northern latitudes of the ACEP has only *Avicennia germinans* present.

The richness of species around the IWP and the decrease in species richness as the distance from the IWP increases in a westerly direction has become known as the mangrove anomaly or the mangrove biogeographic anomaly. The obvious solution is to state that mangroves developed in the IWP and have dispersed out from this region. As discussed earlier, the mangrove fossil record, the relationship between modern and historical mangrove fossils and gastropods, the relationship between available habitat of mangrove and gastropods, and the nestedness between plant and gastropod species all support a Tethys Sea development of mangrove species in the Late Cretaceous Period (Ellison et al. 1999). That is, currently, all fossil evidence opposes an IWP recent origin and dispersal scenario explaining the mangrove anomaly.

The resultant distribution map (Fig. 1.5) is instead likely the result of the early Tethys Seas development followed by more recent vicariance events (Ellison et al. 1999). These vicariance events may be dramatic such as the development of the isthmus of Panama, or merely the flow of ocean currents limiting where propagules can move when detached from the parent, or even ocean temperature extremes creating boundaries the propagules may not be able to cross and remain viable. The relationship between Fig. 1.4 and Fig. 1.5 should not be overlooked in relationship to mangrove species richness, and mangroves species are more numerous in warmer waters.

This manuscript uses the 1960–2007 provincial boundaries of Ecuador before the 2007 redistricting process that saw Santa Elena Province created out of northwestern Guayas Province. Within mainland Ecuador, mangrove forest stands exist in all five of the coastal provinces. As opposed to other global mangrove forests such as those in the IWP that often grow in a narrow fringe directly along the coastline, Ecuadorian mangrove forests are limited to a few secluded coastal estuaries and rarely reside on the exposed coastline. In Ecuador, four interacting conditions likely explain this distribution: (i) the existence of cold-water currents along large portions of Ecuador's coastline; (ii) the high-wave energy along much of Ecuador's coastline; (iii) the lack of a continuous coastal plain along much of Ecuador's coastline; and (iv) the high tidal ranges present along much of this coastline. Each of these four conditions is expanded upon below.

(i) Southern Ecuador, particularly the extension of Guayas into what is now Santa Elena Province, is heavily influenced by the Humboldt Current (Fig. 1.4). The water temperature is regularly below 20 °C and therefore cannot sustain mangrove forests. In Fig. 1.6, this is the extreme southwest of Ecuador in a line approximately from Manta to Santa Elena, and then onto to the western edge of the mangrove forests in the Gulf of Guayaquil. In this region, only the sheltered Gulf of Guayaquil can sustain mangrove forests.
(ii) Secondly, much of the directly exposed Ecuadorian coastline experiences high wave energy from this region of the Pacific Ocean. In addition to the high wave energy, extreme storm events and the episodic appearance of El Niño condi-

Fig. 1.6 Ecuadorian distribution of mangrove species
Map created by the author with 2000 mangrove forest cover data provided by the Giri et al. (2011) and background data provided by Open Street Map and its contributors

tions magnify the wave conditions. Mangrove forests are not suited to high wave energy environments. These high wave energy conditions exist in all but the very northern reaches of the coastline of Ecuador and the sheltered estuaries such as the Gulf of Guayaquil.

(iii) Outside of the few sheltered estuaries, very little if any flat coastal plain exists in Ecuador. Indeed, the transition from the 5000 m peaks of the Andes to the coast is only tens of kilometers in some locations near southern Ecuador. The water then deepens rapidly off the coast with the Ecuador trench dropping to over 3000 m just a few kilometers off much off the Ecuadorian coastline (Graindorge et al. 2004). Such rugged terrain is not well suited to the development of a sizeable forest that prefers vast flat coastal plains found in regions like the Florida Everglades and the Sundarbans of Bangladesh an India. Additionally, such rugged topography, rapid transitions of elevation, and lack of a coastal plain are not suitable for the formation of deltaic environments in which mangrove forests can also become established. The only sizeable deltaic type environments in Ecuador are in the extreme south of the country and the extreme north, both of which have substantial mangrove forests (Fig. 1.6).

(iv) Much of Ecuador has large daily tidal ranges. In the south, Guayaquil has tidal ranges often over 4.5 m. For example, on 1/3/2018, Guayaquil's low tide was −0.4 whereas its high tide was 4.13 m (Flater 2018). Ranges above 3 m are prevalent along most of Ecuador's coastline. Mangrove forests have difficulty becoming established with such large daily tidal ranges. Indeed, this may help explain the near-total dominance of *Rhizophora mangle* in Ecuador as this species is best suited to establishing itself in regions of extreme tidal fluctuation.

Although small mangrove stands are present in all of Ecuador's coastal provinces, the mangrove forest limiting conditions listed above only permit meaningful growth in five distinct and easy-to-delineate areas (Fig. 1.6). The five areas are:

(i) Cayapas-Mataje in Esmeraldas Province in northern Ecuador. This sizeable mangrove forest continues well into Colombia (Fig. 1.6, label CM).
(ii) Muisne estuary in Esmeraldas Province in northern Ecuador (Fig. 1.6, label MS).
(iii) Cojimíes Estuary, which is split between Manabí Province and Esmeraldas Province in northern Ecuador (Fig. 1.6, label CJ).
(iv) Chone Estuary in Manabí Province in central Ecuador (Fig. 1.6, label CE).
(v) The Gulf of Guayaquil in Guayas Province in southern Ecuador which exists along El Oro Province's coastline in southern Ecuador (Fig. 1.6, label GG). This southern mangrove forest area continues into extreme northern Peru.

The five estuarine regions listed above; CM, MS, CJ, CE, and GG (Fig. 1.6), contain greater than 99% of Ecuador's mangrove forests as of 2000 (Giri et al. 2011; Hamilton and Casey 2016) and are all included in these analyses. Some small isolated mangrove stands do exist in other areas along the coast, but none are large than three landsat pixels or 2700 m^2. The isolated single-pixel mangrove stands depicted in Fig. 1.6 are all located on small islands in the mouths of rivers or along river banks. The 2-pixel stands located in the city of Esmeraldas near the mouth of the Esmeraldas River (0.971837, −79.6386905) is mostly on a small island. As of 2017, this small and fragmented forest has received protected status as part of the Sistema Nacional de Áreas Protegidas del Ecuador (SNAP) program (IUCN and UNEP 2016). Another small 2-pixel stand located north of CE and south of CJ is at the mouth of the Jama River on a small island (−0.169535, −80.2652625) in the community of Jama (Fig. 1.6). South of Manta and north of Salinas are a series of small rivers that have a single mangrove forest pixel near the sheltered river mouths of no more than 30 m by 30 m (Fig. 1.6). From north to south, these are small mangrove forest stands in Puerto Cayo at the mouth of the Olina or Jipijapa River (−1.3516763, −80.7349426,224), in Ayampe at the mouth of the Ayampe River (−1.6749111, −80.8095529), and in Palmar in a wetland connected to the Rio Palmar (−2.0201434, −80.7346508) as documented by Shervette et al. (2007).

The author visited all of these isolated mangrove forests between 2007 and 2009 and verified each to be less than 1 ha in size except for the stand at the mouth of the Esmeraldas River which was approximately 1 ha. Other inland mangrove forests depicted are likely small isolated mangrove forest pixels or, more likely, misclassifications. Additionally, no dataset appears to detect a long and narrow strip of mangrove forest that exist along a creek at the southwestern edge of Isla Puná in Guayas Province. This strip of mangrove, although a couple of hundred meters on the north to south axis, is only one or two trees wide along its entire length and hence is likely too narrow to be detected by most remote sensing instruments.

It is likely that historical mangrove forest cover during the last few centuries in Ecuador was more extensive than present, but that all the mangrove cover present was still confined to the same estuaries as today (Fig. 1.6). This location consistency over time is likely due to the fact that mangrove forest location in coastal Ecuador is limited more by geophysical factors than by anthropogenic factors, wheras anthropgenic factors limit the amount of mangrove forest within these locations. That is, the five suitable sites that have large mangrove forests as depicted in Fig. 1.6, likely had more mangrove forests in and around them during pre-history, but no other substantial mangrove forest stands outside of these areas likely existed in modern geological timeframes.

1.4 Mangrove Forest Cover

Since the launch of Landsat in 1972, which heralded the advent of the civilian remote sensing era, mapping of global mangrove forests at high-resolution is possible. Global mangrove forest area estimates pre-1970 are typically back-of-envelope calculations (e.g., as conducted by Boyd and Clay 1998) or extrapolations based on limited input data. Although the era of satellite mapping began with Landsat 1, it is with the launch of the Landsat Thematic Mapper (TM) as part of the Landsat 4 and Landsat 5 systems in the early 1980s, that remote sensing of mangrove forests at the national scale and beyond took hold. No early-period global mangrove forest analysis exists yet from TM, although numerous national and regional analyses were conducted using a variety of different approaches. Indeed, one of the still outstanding mangrove forest research needs is to process historical TM data at 30 m resolution to derive a global mangrove forest estimate for the mid-1980s. A mid-1980s global mangrove forest area data point is needed to provide more meaningful mangrove forest loss information from the global remotely-sensed mangrove forest products that followed in the late 1990s, 2000s, and 2010s. The required 1980s TM global raw data exists at the Global Land Cover Facility (GLCF), but such historical mangrove forest research has yet to be undertaken.

Before and alongside the early remote sensing period, groups such as the FAO, the International Tropical Timber Organization (ITTO), the International Society for Mangrove Ecosystems (ISME) and academics provided global mangrove forest area estimates from a variety of blended sources. Usually, the mangrove forest area calculations were derived from official in-country estimates or the grey literature with various methodologies. Early national estimates should be treated with caution as demonstrated by Mejía-Rentería et al. (2018) in the case of Colombia.

Global mangrove forest cover in 1980 is estimated at 187,940 km^2 (UN FAO 2007) while attempting to account for all mangrove forest holding nations, and as 154,620 km^2 in a 76-country analysis (UN FAO 1981). Saenger et al. (1983) estimate mangrove forest cover as 162,210 km^2 in a 66-country analysis, and by 1985 the FAO was reporting global mangrove forests in a 56-country analysis as covering 165,300 km^2 (UN FAO 2004). Hamilton and Casey (2016) conduct a literature analysis of 1980s global mangrove forest area estimates and establish a global mean value of mangrove forest cover for the 1980s at 167,518 km^2.

Global mangrove forest cover in 1990 was estimated at 169,259 km^2 (UN FAO 2007) while attempting to account for all mangrove forest holding nations, and as high as 198,478 km^2 in 1992 while accounting for only 87-nations (Groombridge 1992). Various estimates for 1993 global mangrove forest area are 141,973 km^2 (ITTO and ISME 1993) and 191,817 km^2 (Fisher and Spalding 1993), in all countries and 87-countries respectively. The first edition of the World Mangrove Atlas, published in 1997, estimates global mangrove forest cover at 181,077 km^2 across 112-countries, which due to the distribution of mangrove forests can be considered a global estimate. Again, Hamilton and Casey (2016) conduct a literature analysis of 1990s global mangrove forest estimates and establish a global mean value of mangrove forest cover for the 1990s at 177,919 km^2.

Although the mean value of global mangrove forest appears to have increased from the 1980s to the 1990s, almost all mangrove forest researchers consider this highly unlikely, and the increase is probably not reflective of ground reality due to three primary methodological reasons. First, more countries are generally included in the 1990s analyses. Second, the countries that are included are surveyed more completely and omit fewer mangrove forest areas. Third, the ability of remote sensing systems to depict ever-smaller patch sizes of mangrove forests increases over time as new and higher-resolution systems come online. All three of these methodological improvements likely lead to more mangrove forest being counted in the 1990s than in the 1980s. One final point to make when comparing the estimates from the 1990s and 1980s is that an averaging approach, or even comparing different estimates against each other within or across a decade, is problematic. Many of the global mangrove forest area estimates are merely aggregates of country estimates that likely consist of many of the same national data points feeding into each global estimate. Therefore, different 1980s estimates and different 1990s estimates are likely to have the same data point feeding into them many times, and some data points likely carry over from the 1980s to the 1990s.

It was with the release of Mangrove Forests of the World (MFW) that a true remotely-sensed global census of mangrove forest area first appeared. Published in 2011, MFW uses 30 m data to estimate mangrove forest cover at the global scale for the year 2000 using a series of 1000 cloud-free Landsat scenes from the GLCF collected in the late 1990s (Giri et al. 2011). It applies a consistent methodological approach globally to mangrove forest mapping and estimates mangrove forest cover at 137,600 km^2 in the year 2000. Since 2011, MFW has acted as a baseline for the numerous other remotely-sensed mangrove forest area estimates that have followed; this fact is evidenced by a current citation count of 1278 for MFW (as of 11/9/18). Other global mangrove forest estimates for 2000 come from a later edition of the World Atlas of Mangroves (Spalding et al. 2010), the UN FAO (2007), Aizpuru et al. (2000), and Hamilton and Casey (2016). They estimate mangrove forest cover at 152,361 km^2, 157,400 km^2, and 83,495 km^2 respectively.[3] One final estimate for the 2000s estimates global mangrove forest cover at 152,310 km^2 for 2005 (UN FAO 2007). Again, Hamilton and Casey (2016) conduct a literature analysis of 2000s global mangrove forest estimates and establish a global mean value of mangrove forest cover for the 2000s at 154,085 km^2.

Finally, for the 2010s, three remotely-sensed estimates of global mangrove forest cover exist. First, for 2010, Bunting et al. (2018) provide a global mangrove forest baseline estimate of 137,600 km^2 in the MDPI journal, Remote Sensing. Hamilton and Casey (2016) estimate global mangrove forest cover in 2012 at 81,849 km^2, but again this is not comparable to other values due to the pixel percentage method utilized and noted above. This 2012 measure was converted to a comparable metric by Hamilton et al. (2018) and resulted in a 2012 global mangrove forest area estimate of approximately of 136,415 km^2.

Figure 1.7 depicts the consensus estimate of decadal mangrove forest area globally. The solid black line is derived from the global mangrove forest area averages compiled by Hamilton and Casey (2016) for the 1980s, 1990s, and 2000s, who derived their mean values from a broad literature review including the grey literature. The 2010s data point is uniquely compiled for this manuscript from all global remotely-sensed mangrove forest estimates for the period post-2010 presented previously; they are averaged in the same manner as the earlier estimates. The dashed

[3] It should be noted that Hamilton and Casey (2016) calculate mangrove forest canopy cover as a percentage pixel cover. This leads to lower estimates than the other pixel presence or absence methodologes employed by all other authors in this list. When adjusted to a presence or absence pixel system, the global mangrove forest area changes for 2000 to 139,158 km^2 (Hamilton et al. 2018).

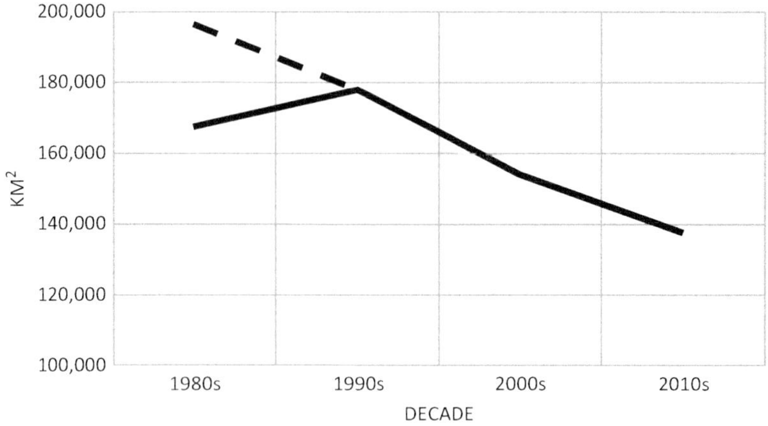

Fig. 1.7 Decadal mangrove forest global area estimates, 1980–2010
The solid line is measured while the dashed line is estimated.

line that extends from 1980 to 1990 is a backward extrapolation from the 2010s, 2000s, and 1990s input data points. This extrapolation was conducted to provide an alternate estimate of average global mangrove forest area value for the 1980s. This 1980s dashed estimate is conducted using a simple linear hindcast based on the data trend from the following decades that have a linear trend. This 1980s hindcast accounts for the issues noted earlier of 1980s global mangrove forest area estimates having only limited countries feed into them, being only limited to a subset of mangrove stands within the countries analyzed, and only having large mangrove forest patch sizes included. The extrapolated hindcast likely provides a more realistic mid-1980s global mangrove forest area value than merely averaging the 1980s individual estimates.

Indonesia currently contains approximately 20–28% of the world's total global mangrove forest area (Table 1.2). Post-1999 remotely-sensed estimates of Indonesia mangrove forest holdings indicate that it contains 20% (Bunting et al. 2018), 23% (Giri et al. 2011), 26% or 28% (Hamilton and Casey 2016) of the global mangrove forest area for the years 2000, 2010, and 2014 respectively. Indonesia has many reasons for its high percentages of mangrove forest holdings. These reasons are:

- Indonesia is entirely contained in the tropical mangrove forest zone (Fig. 1.3).
- The entire Indonesian archipelago has warm ocean currents with no cold water intrusion (Fig. 1.4).
- The majority of Indonesia is centrally located in the IWP region of maximum mangrove species presence (Fig. 1.5).
- Many of the coastlines of the individual islands are isolated and relatively undisturbed by human activity.
- Indonesia contains by far the longest coastline of any country within the tropics. Indeed, at 1° by 1° resolution, Indonesia's coastline is a remarkable 80% of the

Table 1.2 Top-10 mangrove forest holding nations by percent of global mangrove forest holdings

RANK 2000/2010/2014	Country	Percent of global mangrove areas 2000[a]/2010[b]/2014[c]
1/1/1	Indonesia	20/23/28
2/2/2	Brazil	8/8/9
3/3/5	Australia	7/7/4
4/4/6	Mexico	5/7/4
5/5/7	Nigeria	5/5/3
6/6/3	Malaysia	4/4/6
7/7/8	Myanmar	4/4/3
8/8/4	Papua New Guinea	4/4/5
9/9/12	Bangladesh	3/3/2
10/NR/14	Cuba	3/2/2
11/10/23	India	3/2/1
13/NR/9	Venezuela	2/2/3
16/NR/10	Philippines	2/2/3

[a]Siikamäki et al. (2012) reworked national level estimates from Giri et al. (2011)
[b]Bunting et al. (2018)
[c]Hamilton and Casey (2016)

circumference of the earth (Yamanaka et al. 2018). Such a long coastline in the tropics increases the potential number of mangrove forest sites.

Brazil consistently follows Indonesia as the second highest mangrove forest holding nation with 8% or 9% of the global mangrove forest stocks, or approximately one-third of Indonesia's mangrove forest holdings. In addition to Indonesia and Brazil, countries consistently in the top eight mangrove forest holding nations across all databases are Australia, Mexico, Nigeria, Malaysia, Myanmar, and Papua New Guinea. Bangladesh, Cuba, India, Venezuela, and the Philippines take a ninth or tenth spot on the list depending on the exact methodology used, with many countries containing about 1% of the global total once the data extends outside the top 10 mangrove forest holding nations. (Table 1.2).

Across all databases, the top two or three mangrove forest holding countries contain over 35% of the global mangrove forest area, and the top five or six mangrove forest holding countries contain over 53% of the global mangrove forest area (Table 1.2). Of all 13 countries that are inside the top 10 in global mangrove forest area percentage, 12 of them are generally listed as developing nations. Only Australia is listed as having a GDP per capita above the world average in 2017, and Brazil is generally ranked as the next highest GDP per capita of all other nations but is still ranked in 67th place globally (IMF 2017). The dominance of mangrove forest in the developing world likely results in more complex management and tracking of the resource than many other global forest types and a higher potential for unregulated exploitation of the resource.

The historical mangrove forest cover story for Ecuador is opaquer than the global situation. As of 2012, Ecuador is estimated to have between 93,574 ha and

155,957 ha of mangrove forest cover, ranking it as the 20th largest mangrove forest holding globally. (Hamilton et al. 2018; Hamilton and Casey 2016). This estimate is compiled from the Landsat-derived global tree cover data provided in Global Forest Cover (GFC) (Hansen et al. 2013). The lower of these two estimates is a sub-pixel classification of mangrove forest canopy cover, allowing for gaps in the canopy and a hard designation of mangrove species possible presence, whereas the higher number accounts for all trees in the mangrove forest zone, including terrestrial trees located at the aquatic and terrestrial interface. This higher value aligns well with an internal Ecuadorian Satellite Pour l'Observation de la Terre (SPOT) derived estimate of 148,230 ha in 2006 (CLIRSEN 2007). Another Landsat, late-1990s estimate of mangrove forest cover for Ecuador estimated mangrove forests area at 158,261 ha (Giri et al. 2011). Again, this is highly comparable with the other values presented and provides a realistic slightly downward trend between the late 1990s and 2006. The World Mangrove Atlas provides one further estimate. Again, this estimate is primarily Landsat based and estimated Ecuadorian mangrove forest cover at 158,261 ha in 2000 (Rosati et al. 2008; Spalding et al. 2010).

Hamilton et al. (2018) combine all the post-2000 mangrove forest area estimates in Ecuador and convert them into comparable units. In doing so, they provide the best estimate of post-2000 mangrove forest area from all remotely-sensed sources. Their estimate of post-2000 mangrove forest area is 151,887 ha +/− 5% (CI 0.9, n = 5). This estimate is tightly clustered around the mean and seems to have no outliers. That is, all post-2000 unique remotely-sensed mangrove forest cover estimates of Ecuador are in very close alignment. Due to the differing years of each estimate, between 2000 and 2012, this tight clustering indicates relatively little change in Ecuadorian mangrove forest cover post-2000.

Within Ecuador, pre-2000 mangrove forest estimates become scarcer, the methodologies less complete or not reported, and the mangrove forest numbers begin to diverge for the same period. The earliest estimate of mangrove forest area in Ecuador, comparable to this analysis, come from the Ecuadorian Centro de Levantamientos Integrados de Recursos Naturales por Sensores Remotos (CLIRSEN). They report 202,201 ha of mangrove forest cover in 1969 (CLIRSEN 2007). This estimate is compiled using many of the same air photos used in this analysis which were flown in the late-1960s. The methodology is hand digitization of mangrove forest areas from black and white air photographs which are then field verified for delineation accuracy. CLIRSEN even provides a breakdown of the mangrove forest area by province (Table 1.3) that are comparable to the data presented here.

The next date with a mangrove forest cover area estimate for the entirety of Ecuador's coastline is 1979. The 1979 data again ties well to the decadal data presented later in this book. The 1979 estimate of mangrove cover in Ecuador is 203,700 ha (Parks and Bonifaz 1994) (Table 1.3). No methodology is provided, but it is likely that these data are derived from the black and white imagery used in this analysis from 1977. If so, the methodology is hand-digitizing from the black and white photographic plates followed by field verification. It is possible that mangrove forest area remained stable from the late 1960s to the late 1970s as shrimp farms, a

Table 1.3 Ecuadorian mangrove forest cover by year and province when available, 1969–1999

Year/Hectare	Esmeraldas	Manabí	Guayas	El Oro	Total	Author
1969	32,343	12,099	122,615	35,144	202,201	CLIRSEN (2007)
1979					203,700	Parks and Bonifaz (1994)
1980					203,000	UN FAO (2004)
1984					182,100	Parks and Bonifaz (1994)
1987	29,257	6401	116,065	23,403	175,126	CLIRSEN (1987)
1987					237,700[a]	Spalding et al. (1997)
1990					163,000	(UN FAO 2004)
1991					196,000	UN FAO (2004)
1991	23,969	6953	109,928	20,918	161,768	Bodero (1993)
1991					177,600	Harcourt and Sayer (1996)
1992			109,000		246,900[b]	Spalding et al. (1997)
1999	23,189	1797	104,586	18,911	148,483	CLIRSEN (2007)
1999					149,556	UN FAO (2007)

[a]In Table X3, both mangrove forest estimates presented, but not constructed, by Spalding et al. (1997) should be treated with a high level of uncertainty as they are contradicted by the analyses in this book, the analysis by (CLIRSEN 2007), the analysis by Parks and Bonifaz (1994), and the analysis by (Bodero 1993), which are all others in close agreement. Additionally, these two high and outlier estimates provide little or no meaningful methodology

[b]As above

significant contributor to mangrove forest decline, are still in a nascent stage at this time (Parks and Bonifaz 1994), and the mangrove forests were still recovering from earlier tannin-driven deforestation.

By the 1980s, mangrove forest estimates become more common, with at least three mangrove forest area estimates produced. The two reliable estimates are again from CLIRSEN directly or from CLIRSEN via other literature sources. The CLIRSEN direct estimate uses a synthesis of Landsat Multispectral Scanner (MSS), SPOT, 1:60,000 aerial photography, and 1:25,000 topographic maps to delineate mangrove forests. The topographic maps are created from aerial photography and field verification. The Parks and Bonifaz (1994) methodology is undefined, but they do state they obtain their data from CLIRSEN, so the methods are likely consistent. The two CLIRSEN-derived estimates are 182,100 ha (CLIRSEN 1987) and 175,100 ha (Parks and Bonifaz 1994) (Table 1.3). Again, this loss from the late 1960s and late 1970s levels of approximately 200,000 ha of mangrove forest appears credible as shrimp aquaculture was rapidly expanding during this period in Ecuador (UN FAO Fisheries and Aquaculture Department 2017), often at the expense of mangrove forest (Hamilton and Stankwitz 2012; Hamilton 2013). The 237,700 ha estimate for 1987 (Table 1.3) provided by Spalding et al. (1997) via Harcourt and Sayer (1996) should likely be discarded as it is not a credible estimate and has no methodology provided. Indeed, the authors note when discussing this estimate that it is only used due to the absence of other data and without further details, it is difficult to assess its accuracy (Spalding et al. 1997).

By the 1990s, mangrove forest cover estimates for Ecuador had become more numerous with four estimates, but they are often still lacking well-defined methodologies. The 1980s mangrove forest area estimates, aside from one odd outlier, continue their downward progression (Table 1.3). The two 1991 estimates of 161,768 ha (Bodero 1993) and 177,600 ha (Harcourt and Sayer 1996) ha respectively seem credible, although the lower figure more closely fits the established trendline (Table 1.3). Again, both estimates omit robust and well-defined methodologies. As with their 1987 estimate, the 1992 estimate by (Spalding et al. 1997) is best discarded as it is a substantial outlier that goes against the other mangrove forest area estimates as presented by all other spatiotemporal analyses, and it again lacks a robust or well-defined methodology. The final estimate for the 1990s presented by CLIRSEN (2007) points to a mangrove forest area of 148,483 ha (CLIRSEN 2007) in 1999. The 1999 mangrove forest area presented by CLIRSEN again fits well into the trendline of declining mangrove forest area from the 1970s onwards.

Figure 1.8 depicts mangrove forest levels in Ecuador since 1969 and is compiled with the data from Table 1.3 using the Ecuadorian CLIRSEN or CLIRSEN-derived data sources. Between 1969 and 1979 mangrove forest area appears to have remained mostly stable. From 1979 to 1999, 55,217 ha of mangrove forest appears to have been lost from the coastline of Ecuador. This loss represents 27% of all Ecuadorian mangrove forest cover in 20 years. This level of loss equates to a 1.35% annual loss of mangrove forest during this period along the coast of Ecuador. To give this 27% mangrove forest loss value perspective, the much-publicized Brazilian rainforest loss is estimated to be 10% for the 20 years from 1977 to 1997, and this equates to 0.5% annually (Butler 2018). That is, during the 1980s and 1990s mangrove forest loss in Ecuador was proceeding at a rate that was likely 2.7 times faster than the loss rate of the Brazilian Amazon rainforest.

The data provided by CLIRSEN allows for an estimate of mangrove forest losses by individual province to be conducted. Although spatially explicit, not all time-

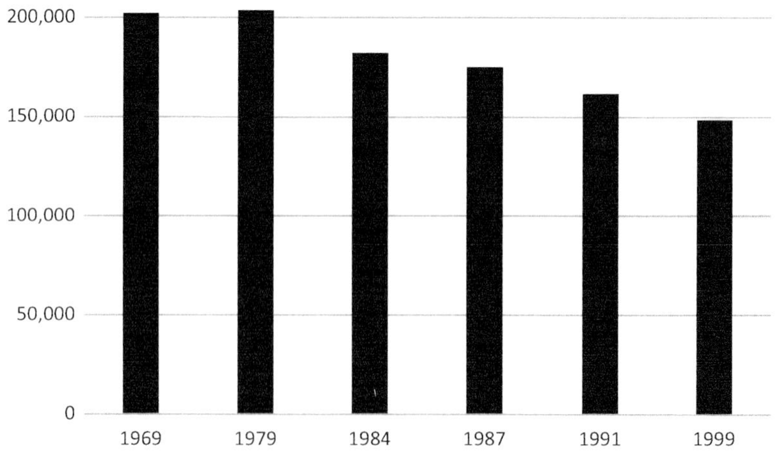

Fig. 1.8 Ecuadorian mangrove forest area estimates in hectare, 1969–1999

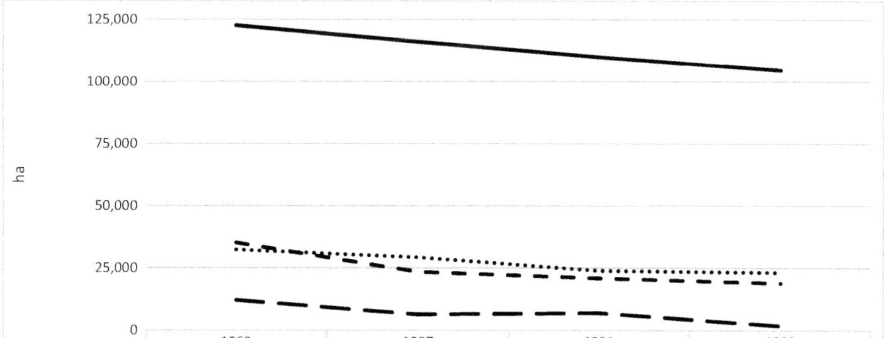

	1969	1987	1991	1999
Esmeraldas	32,343	29,257	23,969	23,189
Manabí	12,099	6,401	6,953	1,797
Guayas	122,615	116,065	109,928	104,586
El Oro	35,144	23,403	20,918	18,911

Fig. 1.9 Ecuadorian mangrove forest area by year and province, 1969–1999

periods are covered. Between 1969 and 1999, Esmeraldas Province saw its mangrove forest area decline from 32,343 ha to 23,189 ha and this equates to a 26% loss over these 30 years (Table 1.3, Fig. 1.9). This period of loss equates to an annual loss rate of 0.87% over the entire 30-year period. The loss in Esmeraldas appears to be concentrated in one small timeframe, the 4 years between 1987 and 1991 (Table 1.3). The mangrove forest loss during these 4 years is 5288 ha or 18% of the 1987 mangrove forest area. This four-year mangrove forest loss rate is 4.5% annually.

Manabí Province appears to have suffered dramatic mangrove forest losses between 1969 and 1999. The mangrove forest baseline in 1969 is 12,099 ha, but by 1999, this had decreased to 1797 ha (Fig. 1.9). Mangrove forests in Manabí Province lost 85% of their total area in only 30 years. This is a remarkable loss rate of 2.8% annually for the entire 30-year period. Such an annual loss rate is unlikely to be matched anywhere globally across almost any forest type. The mangrove forest losses are high across all periods within Manabí.

Guayas Province appears to be the least impacted of all provinces. During the 30-year analysis period, Guayas lost 14.7% of its mangrove forest (Fig. 1.9). Guayas is the only province that appears to have an annual loss rate far below 1%, with an annual loss rate of 0.5%. The substantially lower loss rate is important because in 1969 Guayas Province contained almost 61% of Ecuador's mangrove forest (Fig. 1.9). Although it has a far lower loss rate than any other estuary, Guayas lost 18,029 ha of mangrove forest during this period at a loss rate that still far outstrips Brazilian Amazon rainforest loss during this period. Unlike Esmeraldas, but like Manabí, the loss rates appear entirely consistent over the 30 years.

El Oro Province lost 46%, or 16,233 ha, of its mangrove forest between 1969 and 1999 (Fig. 1.9). This equates to an annual loss rate of 1.5% annually. Most of the mangrove forest losses in El Oro Province appear to have occurred before 1991, with a substantial slowing of the mangrove forest loss rate between 1991 and 1999 (Fig. 1.9). Again, this is a remarkable loss-rate to be sustained for this length of time.

Ecuadorian mangrove forest levels, be them national or provincial, were more difficult to determine before the onset of the satellite era and the preceding and air photo era, and much of the information is likely lost to history. Between 1940 and 1970 the academic literature documents significant clearing of mangrove forests for the tannin industry, a major commercial extraction industry in Ecuador during that period (Hamilton and Collins 2013; Snedaker 1986; Labastida 1995) and local residents confirm this clearing. Indeed, much of the Ecuadorian mangrove forest now present is likely regrowth from this earlier, almost total clearing for tannin extraction, particularly in Ecuador's northern estuaries (Collins 2010; Ocampo-Thomason 2006). By the 1970s, the industry had substantially declined due to the collapse of the global tannin industry and the switch to terrestrial sources of tannin (Snedaker 1986).

Mangrove forest exploitation in Ecuador is not unique to the twentieth, and twenty-first centuries. During the nineteenth, eighteenth, seventeenth and even the sixteenth centuries, mangrove forest exploitation is documented for Ecuador. During the early period of Spanish colonization in the sixteenth Century, mangrove forest resources in Ecuador were an important part of the colonial economy. The Spanish utilized Ecuadorian mangrove forest timber for construction, shipbuilding, leatherwork (tannins), and as charcoal in the sugar production industry; during this period mangrove forest wood was a part of the tribute paid by indigenous communities to the Spanish king (López-Angarita et al. 2016). It is likely that indigenous communities exploited mangrove forests long before the Spanish arrival as well, with family groups using them for food, timber, and fuel, and specialists using them for commercial and religious products (Bodero 1993).

By the seventeenth century, the Spanish were building ships from mangrove forest timber in Guayaquil (López-Angarita et al. 2016). In the seventeenth and eighteenth centuries, mangrove forest wood was exported from Guayas and Esmeraldas to help build cities such as Lima. The demand for Ecuadorian mangrove forest wood was so high that the Spanish crown was forced to regulate the industry via licenses and permits (López-Angarita et al. 2016). Finally, into the nineteenth and early twentieth century periods of Ecuadorian independence, mangrove forest wood was exploited without regard to its sustainability by the newly formed republic that was strapped for foreign currency and had little or no experience or understanding of the importance of managing the critical mangrove forest resource (López-Angarita et al. 2016).

None of the above studies between 1969 and 2012 answer the sub-province, estuary by estuary mangrove forest loss question, nor do they paint a broader picture of what was occurring in the estuaries that drove these substantial declines in post-1970 mangrove forest cover. The issue of Ecuadorian mangrove deforestation has not garnered the attention gained by other tropical transitions such as loss of the nearby Amazon rainforest, or even the loss of nearby coal habitats. Important questions remain unanswered, namely: What was replacing the mangrove forest and why? How did the loss occur both spatially and temporally? Who are the winners and losers from this transition? This manuscript not only attempts to document the mangrove forest loss that occurred at a granularity far below the estuary level, but it also attempts to answer some of these more related but complex questions.

1.5 Mangroves as a Keystone Species

Before the 1970s, the mainstream academic community remained mostly unconcerned with the ecological role of mangrove forests, whereas the wider public and policy community often viewed mangrove forests as an unimportant or nuisance landcover. Within the academic community, and later among the wider public, this negative view of mangrove forests began to change, in part due to the pioneering work of Odum and Heald (1972). Odum and Heald meticulously examined the role of mangrove forests in South Florida in supporting not only the entire coastal South Florida ecosystem but additionally powering the entire Gulf of Mexico fishery by supporting a wide diversity of species including large finfish and the commercially important shrimp fishery. Arising from this pioneering research among the mangrove forests of South Florida, a group of scientific researchers interested in studying mangrove forests, particularly their broader ecosystems role, began to expand the depth and breadth of mangrove forest scientific research. It is from the pioneering work of Odum and Head, and those that followed, that the view of mangrove forest changed from a nuisance or unimportant landcover to a keystone mutualist that supports the ecosystem of the region in which they reside.

Before 1970 and the research mentioned above, mangrove forests were generally only valued for what they could be converted into (Ellison and Farnsworth 1996). Lugo and Snedaker (1974) state that mangrove forests were most often viewed as a landcover with little or no value to society and the United States Department of Agriculture Soil Conservation Service did not classify mangrove forests as having any use in agriculture or forestry (Lugo and Snedaker 1974). As late as 2000, academic concern over the loss of mangrove forest due to the expansion of aquaculture could only be found in highly specialized academic journals and not in the broader environmental literature (Valiela et al. 2001). To this day, the public and policymakers generally do not have the same understanding of mangrove forest loss and its environmental significance and extent that other tropical habitats such as tropical rainforest and coral reefs enjoy.

The negative or entirely omitted view of mangrove forests is in retreat, and the importance of mangrove forests is now apparent within the estuarine research community and beyond. Mangrove forests are now viewed as one of the most productive ecosystems on the planet (Costanza et al. 1997; Costanza et al. 2014). Indeed, mangroves are now often viewed as the keystone coastal species because not only are they highly productive in and of themselves, they create the ecosystem in which they reside by stabilizing the soil and providing habitat in which other organisms flourish (Hogarth 2007). Thus, mangrove forests actually help establish and then support one of the most biologically diverse ecosystems on earth (Kathiresan and Qasim 2005) in what would otherwise be quite barren regions. For example, on the Caribbean Islands and along the coastlines of Colombia, mangrove forests support over 140 bird species, 200 fish species, and many hundreds of terrestrial and marine invertebrates, and they are the basis for the high floral and faunal biodiversity in otherwise low biodiversity mud and salt flats (Alvarez-León and Garcia-Hansen 2003). Mangrove forests play the same role in Ecuador, perhaps with even more significance regarding biodiversity, as coastal Ecuador has undergone a dramatic extinction of flora and fauna, and these extinctions are driven by mangrove deforestation (Dodson and Gentry 1991). Despite massive mangrove deforestation across most of Ecuador's estuaries, these estuarine environments have mostly maintained the same relatively high biodiversity associated with other global less-impacted mangrove forests.

1.6 Mangrove Forests and the Support of Wild Fisheries

Table 1.4 describes the significant functions of mangrove forests, with reference to local coastal populations. The indirect food support played by mangrove forests are mostly related to the support of fisheries from crustaceans through to larger finfish.

Table 1.4 Traditional mangrove forest goods and services (Hamilton and Collins 2013)

Direct Food	Timber	Mitigation	Habitat	Other
Wild shrimp	Charcoal	River flood control and protection	Maintenance of biodiversity	Tourism
Wild fish	Firewood	Shoreline stabilization	Fish Hatchery	Recreation
Baitfish	Boats	Wind protection	Juvenile habitat	Medicinal plants
Mollusk/Clam	Poles	Water purification	Migratory Bird Habitat	
Crab	Home construction	Wastewater treatment	Coral habitat support	
Eel	Thatched roofing	Carbon sequestration	Pollinating bats and bees	
Traditional aquaculture products	Tannins/dyes	Groundwater management		
Cockles	Fish Smoking	Pollutant treatment (agriculture runoff)		
Edible bark		Ocean/surge protection		
Edible plants				
Pollinating species				
Sugar				
Honey				
Alcohol				

Note: Table adapted from various sources (Macnae and Kalk 1962; Macnae 1968; Tomlinson 1986; Blanchard and Prado 1995; Naylor et al. 1998; Armitage 2002; Batagoda 2003; Warne 2007; UN FAO 2004; Conchedda et al. 2011)

Numerous commercially important fish species use mangrove forest as a nursery or as a location to avoid larger predators, and artisanal fishers frequent the waters around mangrove forests across the tropical word. It is not just the support of off-shore or even estuarine fisheries that is important, certain mollusks, such as the commercially important *Anadara tuberculosa* (commonly called cockles or locally, concha negra) in Ecuador, rely entirely on mangrove forest for their habitat and survival (Borda and Cruz 2004; Moreno and Mora 2009). Without mangrove forests, these bivalves have no habitat. Indeed, mangrove forest supports direct local seafood consumption in the form of crab and shrimp (Armitage 2002) in addition to mollusks and fish.

Mangrove forests sustain many livelihood uses, which are expanded upon below, but the most important role of mangrove forests is likely their essential support of fisheries. It is known that commercial fisheries around the world are overexploited, and it is estimated that 93% of global marine fish stocks are either fully fished or overexploited (UN FAO Fisheries and Aquaculture Department 2018). Mangrove forests are essential to commercial fisheries as they provide both food and shelter to wild fish and invertebrates (Hutchison et al. 2014). The fisheries support role is more nuanced than merely providing direct habitat and shelter. In a review of the role of mangroves in supporting fisheries, Hutchison et al. (2014) state that mangroves support fisheries via the following mechanisms:

- The detritus and leaves expunged by mangrove forests are a critical part of the food chain, supporting many fisheries via decomposers.
- Fisheries are supported by the primary productivity of both periphyton and phytoplankton that occur in mangrove forests and form the base of the food web on which the exploited fish and invertebrates depend.
- Mangrove forests capture incoming nutrients, process them, then export them in the form of organic carbon and living biomass, such as planktonic larvae and immature fish and invertebrates.
- "*Species of interest to the fisheries sector are found at all levels of the food chain, with detritivores such as crabs, prawns, and mullet; filter-feeding bivalves, planktivorous fish such as herring and anchovy species, and higher consumers such as some mud crabs and much other fish including snappers and groupers.*" (Hutchison et al. 2014 p. 6)
- Many marine species use mangrove forests as nurseries. These include species that spend their formative years in the mangrove forest before moving to habitats farther offshore such as coral reefs. Thus, these non-mangrove forest habitats have their stocks renourished via mangrove forests.

Quantifying this support of fisheries is difficult, but research has been undertaken. For example, the mangrove forests of Malaysia are estimated to support 50% of Malaysia's entire coastal fishery by providing trophic and refuge support, as well as larval retention. The value of Malaysia's mangrove forests to fishers alone was approximately $70 million in 1999 (Chong 2007; Ebarvia and Corazón 1999). In Florida, mangrove forests offer habitat, shelter, and food sources for species at the base of the food chain (Gore 1977; Odum and Heald 1972). It is estimated that 90% of fish species along Florida's coastline depend on mangrove forests, meaning that to exist, these fishes rely on the mangrove forest at some point during their life cycles (Gore 1977). This pattern is recognized worldwide. In Sulawesi, Indonesia, mangrove forest provides habitat for the most commercially important marine species in the region, including species of fish, shrimp, crab, and mollusks (Armitage 2002). Additional evidence for mangrove forests supporting fisheries are provided by Chong (2007); Frias-Torres et al. (2007); Granek and Frasier (2007); Koenig et al. (2007); Lugendo et al. (2007); Nagelkerken (2007); and Hutchison et al. (2014).

Although controversial and now mostly disproven, there are arguments against the importance of mangroves in supporting wild offshore fisheries. It is stated that the majority of studies show that mangrove forest losses correlate to fisheries decline, but that the data provided does not show these losses as the cause of fisheries decline (Blaber 2007). A more reasonable explanation, argues Blaber (2007), is that the correlation between mangrove forest depletion and overfishing on a similar spatiotemporal scale, and that the relationship between the decline in commercial offshore fisheries resulting from the decline in mangrove forests is dogma. Although questioned at the offshore level, little doubt remains about the linkage between artisanal nearshore fisheries and mangrove forests. Even (Blaber 2007 p. 465), states that:

> It is important to distinguish between 'fisheries within mangrove systems,' usually of an artisanal or subsistence nature in developing countries, and 'offshore (of mangroves) fisheries' that are usually commercial or industrial concerns. In the former case, the activities by traditional or artisanal fishermen may be long-established and are totally dependent on the existence of the mangrove system.

Artisanal communities have raised fish and shrimp for hundreds of years in mangrove forests by building small mud levies in the forests and returning at a later date to harvest the species captured within the levees (Macnae and Kalk 1962; Naylor et al. 1998). Additional direct food resources including exploiting mangrove forests for sugar, honey, edible plants, and even production of alcohol. Indirectly, mangrove litter has been used as feed, particularly for goats and cattle in traditional communities (Tomascik 1997), again providing food or livelihood options to local communities. Another vital role of mangrove forest is as habitat for bats and bees which are pollinators of many coastal regions' essential cash crops.

Ecuador lacks a robust set of peer-reviewed studies examining the relationship between mangrove forests and commercially important offshore fish species. There is one study comparing fish populations offshore of mangrove forest at the mouth of the Rio Palmar to offshore of the mangrove-free river mouth of Rio Javita, a river approximately 1.5 miles from the mouth of the first river (Shervette et al. 2007). This study identified 36 fish species representing sixteen families within the Rio Palmar and Rio Javita. Twenty-one of these species occurred only in Rio Palmar which retains a limited mangrove forest, as opposed to the cleared Rio Javita, which now contains no mangrove forest. Species diversity and richness were higher in the mangrove forest area, and more importantly, nine of the twelve economically important fish species occurred only offshore of the river with a mangrove forest (Shervette et al. 2007) despite the much-altered and reduced state of this remaining mangrove forest.

The comparative analysis conducted by Shervette et al. (2007) is even more startling as it was conducted in an intertidal environment which had been almost entirely stripped of the majority of its former mangrove forest. Landsat imagery interpretation of the area showed the intertidal mouth of the Rio Palmar had 20-fold more area under aquaculture in 1973 than the remaining area of mangrove forest in 2007. That is, even though only a small portion of Rio Palmar's mangrove forest remains intact, it still appears to sustain more fish species, and more commercially important fish species, than the nearby Rio Javita that has suffered total deforestation of its former mangroves.

Within Ecuador, mangrove forest plays a vital role in the habitat and harvesting of the lesser-known fishery for the concha negra (*Anadara tuberculosa* and *Anadara similis*, cockle in English). Concha negra is a black-shelled Bivalvia that appears to only reside in limited sites along the Pacific coastline of the Americas (including Ecuador) and resides only in the root structure of *Rhizophora mangle* (Palomares and Pauly 2018; Cárdenas et al. 2018). Concha negra is wholly reliant on the mangrove forest for its survival. Put simply, without substantial mangrove forest cover and an intact root sturcure, concha negra and the native livelihoods that depend on it cannot exist. Concha negra is considered a delicacy, used primarily in Ecuadorian ceviche, and its population is stressed (Cárdenas et al. 2018) to the point of potentially becoming extinct (Acosta 2018). Within Ecuador, a minimal and non-harvestable population still exists on the El Oro and Guayas border, as well as an extensive fishery in Reserva Ecológica Manglares Cayapas Mataje (REMACAN) on the Colombian border. In and around REMACAN, it has been shown that 85% of the traditional residents in this area rely on fishing or harvesting of concha negra for their income, and 10% on the concha negra alone (Ocampo-Thomason 2006).

Concha negra additionally plays a vital role in gender dynamics in northern Ecuadorian fisheries. Concha negra collection in and around REMACAN is an almost wholly female activity (Ocampo-Thomason 2006; Veach 1996), with 82.4% of gatherers being female (Ocampo-Thomason 2006). Concha negra, particularly *Anadara tuberculosa,* is in short supply and fetches a high price, enough to provide a substantial household income to the female cockle gatherers (Ocampo-Thomason 2006; Veach 1996). The women, known as concheras, appear to manage the harvest sustainably by closing grounds for many years until they are returned to the harvesting cycle for a short period before being closed again. Indeed, the concheras appear to be driving the resistance to deforestation in and around REMACAN for the last decade. Within Ecuador, and much of Latin America, each town has a statue of an important historical figure that relates to the town. In San Lorenzo near REMACAN, this statue consists partly of an Afro-Ecuadorian female collecting concha negra in a woven basket.

1.7 Traditional Uses and the Economic of Value of Mangrove Forests

Following direct and indirect food supply, mangrove forests next most important use is likely for timber and timber products (Table 1.4). Mangrove forests are harvested for charcoal, for direct use as firewood, and other timber-related use such as the construction of homes and poles for fishing (Macnae 1968; Tomlinson 1986; Naylor et al. 1998; Armitage 2002; Warne 2007; UN FAO 2004). For example, mangrove forest wood is used, in a potentially unsustainable manner, for fish smoking in Senegal (Conchedda et al. 2011). The exploitation of mangrove forest tannins for use in the leather industry and beyond has existed in Latin America from colonial times and reached industrial levels by the 1950s and 1960s before declining in the 1970s (López-Angarita et al. 2016).

Besides their economic importance as an extractive, mangrove forests play an essential mitigation role, particularly as it pertains to storm defense (Table 1.4). This role often directly benefits local communities living alongside the mangrove forests. For example, mangrove forests reduce wave energy thus limiting damage to local communities from large storms; they reduce tsunami damage by breaking and reducing the energy of incoming waves; they reduce erosion by buffering wave energy, and mangrove forests may even mitigate sea-level rise (Spalding et al. 2014). The role of mangrove forests in mitigation is likely one of the less accounted for and advocated roles of mangrove forests globally.

Ecuador has the same patterns of mangrove forest exploitation that are common globally, with use for fuel, timber, and foodstuffs dominating. *Rhizophora mangle*, which is by far the most dominant mangrove species in Ecuador, is particularly suited to charcoal and for use as a heat or cooking sources. For example, it has been shown that *Rhizophora mangle* has a heat value of 4.400 kcal/kg in Ecuador, making it a high-density heat source (Doat 1977). For comparison to a terrestrial forest, *Rhizophora mangle* has 48% more energy per unit area than an equivalent pine forest with the same standing volume (UN FAO 2004). It is particularly useful for charcoal due to this heat potential, but an additional benefit is the relatively low moisture content when compared to other mangroves. *Rhizophora mangle* has a felled moisture content of approximately 40%, whereas *Avicennia germinans* has a moisture content closer to 95% and therefore takes three times as long to dry as *Rhizophora mangle* (UN FAO 2004). The lower moisture content of *Rhizophora mangle* makes it more suitable for use as a heat source, for cooking, and conversion into charcoal.

In addition to its use as a heat and fuel source, *Rhizophora mangle* is particularly suitable in construction for use in the aquatic environment such as in boats, poles, and home construction in regularly flooded environments. For example, in and around San Lorenzo, Ecuador, and much of artisanal coastal Ecuador, homes are built on wooden stilts in an environment that floods almost daily at high tide or is permanently in the water. These homes used to be built from mangrove forest timber, but the current use of mangroves for this purpose appears more limited

(Ocampo-Thomason 2006). The exploitation of mangrove forests for timber used for boat construction has been known since at least colonial times, with Guayaquil becoming a mangrove-based ship construction center during Ecuador's Spanish colonial period (López-Angarita et al. 2016). While current use for this purpose is likely more limited with fiberglass boats now dominating (Ocampo-Thomason 2006), some fisherfolk still used utilize mangrove timber in boat construction as recently as the late-1990s (Veach 1996).

Outside of fisheries support and timber or wood uses, mangrove forests provide a host of other lesser-known livelihood opportunities in Ecuador. Hamilton and Collins (2013) indicate that the people of central and northern Ecuador still use the mangroves medicinally as a coagulant and for shoe manufacturing. Mangrove forests appear to play an important tourism role in both REMACAN and Chone, with artisanal fishers giving regular tours of the mangrove forest. US-sponsored not-for-profit research or educational centers that focus on the mangrove forest exist in Guayas, Chone, Muisne, and near REMACAN and likely provide a substantial stimulus to the local economy. Of course, all these activities rely on healthy mangrove forests.

Placing economic values on a preserved and functioning natural ecosystem is problematic for a few reasons. The economic returns of ecosystems are challenging to measure as they are often provided over long multigenerational periods, often distributed in small amounts to many people including those that may reside far from the ecosystem, and many of the returns of mangrove forests ecosystems are intangible and therefore difficult to monetize. Despite all these issues, environmental economists still attempt to undertake such valuations. Mangrove forests and the estuaries in which they reside provide some of the highest economic returns on the planet (Costanza et al. 1997, 2014). The rate and extent of global mangrove deforestation and the associated estuarine disturbances are in part due to the slow realization of these ecosystems' economic values by economists (Blaber 2007).

For all the reasons listed above, mangrove forests act as an essential local economic engine to populations that reside near the forests, they produce valuable economic outputs from the sun and sea, then pass on these goods to the local communities in many forms (Tomlinson 1986).

1.8 Mangrove Forest Carbon Stocks and Sequestration

In the last 5–10 years, the role of mangrove forests in climate change has come to the forefront, and this new research frontier likely increases the economic value of mangrove forests multiple times over and may be a driving force in their future protection. It has been known for some time that mangroves sequester atmospheric CO_2 and convert this CO_2 into buried carbon. For example, as early as 2002 it was noted that mangrove forest sediments could contain as much 700 tonnes of carbon per hectare (1 m deep) and that mangroves can sequester some 1.5 tonnes of carbon per hectare per year (Ong 2002). Since this time, the study of the mangrove forest carbon sector has become a robust sub-discipline within the mangrove forest scientific community.

Globally, mangrove forests contain approximately 4.19 Pg of carbon and Indonesia, Brazil, Malaysia, and Papua New Guinea account for greater than 50% of the global mangrove forest carbon stocks (Hamilton and Friess 2018). Of this 4.19 Pg, 2.96 Pg is in the sediment and 1.23 Pg in the living mangrove trees. Between 2000 and 2012, mangrove deforestation resulted in a maximum potential of 316,996,250 tonnes of CO_2 equivalent emissions (Hamilton and Friess 2018). Mangrove forests contain significant mangrove carbon stocks; indeed mangrove forests may have the highest carbon stocks per hectare of any forest type globally, containing on average 1023 Mg of carbon per hectare. This carbon stock estimate is at least three times more than either boreal forest, temperate forest, or tropical forests (Donato et al. 2011). Indeed, it has been proposed that for a specific group of tropical nations with emerging economies and large mangrove forest areas, mangrove forest carbon sequestration can help meet the rigorous carbon targets set out as part of the Paris Agreement on Climate Change by offsetting national emissions (Taillardat et al. 2018).

Ecuadorian mangrove forests may be regionally important carbon sinks as they consist primarily of *Rhizophora mangle*, which likely contains some of the highest carbon stocks of any mangroves globally due to its extreme potential height and substantial trunk diameter. It is estimated that Ecuador's mangrove forests contain 57,467,199 t C or 1.36% of the entire global mangrove forest carbon stock as of 2012 (Hamilton and Friess 2018). This 1.36% is likely an underestimate as it does not account for the fact that Ecuador's mangrove forests are almost entirely dominated by *Rhizophora mangle*. In an analysis of Ecuador's mangrove forests (excluding much of Guayas) using a suite of models that account for the field-verified Ecuadorian mangrove species distribution, it is estimated that the northern forests of Ecuador contained 18,754,752 t C ± 27% of living biomass (Hamilton and Lovette 2015). As it has been established that the mangrove forests of Esmeraldas, Manabí, and Elo Oro Provinces contain approximately 30% of the Ecuadorian mangrove forest area (Table 1.3), it can be inferred that countrywide Ecuadorian mangrove forest carbon stocks are 3.3 times higher than the non-Guayas value, or approximately 62,515,840 t C would be contained in the living biomass of Ecuador's mangrove forests. As biomass typically only accounts for 41% of the total carbon in the mangrove forest system, the other 69% being in the soil (this is a conservative estimate), a whole-system assessment can be compiled.

It can be quickly estimated that using Ecuador-specific parameters and conservative estimates of soil depth and the ratio between living biomass carbon and sediment carbon, that Ecuador's mangrove forests contain approximately 152,477,659 t C. This equates to Ecuador having the sixth highest mangrove forest carbon stocks globally if incorporated into the most recent global mangrove forest carbon stocks database (Hamilton and Friess 2018). It should be noted that Colombia, and to a lesser extent Mexico, would likely also increase by a similar ratio using this site-specific methodology that accounted for *Rhizophora mangle*, meaning Ecuador would drop to about eighth on the global list.

1.9 Mangrove Forest Summary

Mangrove forests, and the estuaries in which they reside, are undoubtedly one of the most valuable ecosystems on the planet due to their support of wild fisheries, their raw economic valuation from the timber resource, their carbon sequestration potential, the livelihoods they provide, their essential environmental mitigation functions, and their role in artisanal and wider food security. Mangrove forests benefit the world by sequestering high amounts of CO_2, driving many commercially important nearshore and offshore fisheries, and providing high levels of biodiversity in what are otherwise often species-limited environments. Historically, they have provided a globally important source of timber for poles and home construction, tannin for the leather industry, and numerous commercially important fish species. All of these have been extracted from mangrove forests for consumption, often far away from the mangrove forest location.

Mangrove forests provide numerous economic opportunities to local populations who reside close to the forests. They not only provide income to residents but have historically provided a secure supply of protein during times of food stress. Local residents harvest the cockles and catch the wild shrimp that are both wholly reliant on the mangrove forests for their survival. Artisanal fisherfolk have traditionally built their homes and boats from mangrove forest timber and used the mangrove wood for heat and cooking with charcoal. The mangrove forest houses bats, insects, and bees that pollinate the local crops of the farmers near the estuaries. These artisanal communities are physically protected by the mitigating function of mangrove forests that limit the impacts of hurricanes, El Niño events, storms, erosion, and even sea level rise. Within Ecuador and beyond, it is the local populations that benefit the most from the mangrove forests and are therefore most impacted by their removal.

As can be seen from the mangrove forests levels presented in Table 1.3, Figs. 1.8 and 1.9, Ecuadorian mangrove forest is under threat. The arrival of shrimp aquaculture to this region in the 1970s still reverberates throughout Ecuador's mangrove forests and the communities that have historically exploited as well as conserved them.

1.10 Recommended Further Reading

Mangrove forest Adaptations to Saline and Anaerobic Environments

Kelvin et al. (2001)

Mangrove forest Biology and Mangrove forest Botany

Hogarth (2015); Tomlinson (1986)

Mangroves and Wild Fisheries

Bulletin of Marine Science, Volume 80, Number 3, May 2007
Proceedings of the First International Symposium on Mangroves as Fish Habitat.

Traditional Use of mangrove

Macnae and Kalk (1962); UN FAO (2004)

The Mangrove forest Diversity Anomaly and the Prehistory of Mangroves

Ellison et al. (1999)

Ecuadorian Mangrove species

Guevara and Granda (2009)

Global Mangrove forest Levels

Hamilton and Casey (2016); Giri et al. (2011)

Global Mangrove forest Carbon Stocks

Donato et al. (2011); Hamilton and Friess (2018)

Ecuadorian Mangrove forest Levels

Parks and Bonifaz (1994)

Ecuadorian Mangrove forest Carbon Stocks

Hamilton et al. (2017); Hamilton and Lovette (2015)

Concheros in Ecuador and the Local Mangrove forest Economy

Ocampo-Thomason (2006)

General Mangrove forest Background

Spalding et al. (2010)

1.11 Data Availability

Figure 1.3, Global Distribution of Mangrove, circa 2000
Mangroves Forests of the World (MWF)
Data – http://sedac.ciesin.columbia.edu/data/set/lulc-global-mangrove-forests-distribution-2000

Figure 1.4, Global Ocean Currents and Mangrove forest Distribution
Data – https://doi.org/10.7910/DVN/TKGO2Z
Tutorial video – https://youtu.be/nN8sl7RuXXs

Figure 1.5, Global Distribution of Mangrove species
Data – https://doi.org/10.7910/DVN/FK8JZK

References

Acosta JRS (2018) Chain of supply of the ecuadorian black shell and its risk of extinction. Janus. http://janus.ec/cadena-de-suministro-de-la-concha-negra-ecuatoriana-y-su-riesgo-de-extincion/. Accessed 17 Dec 2018

Aizpuru M, Achard F, Blasco F (2000) Global assessment of cover change of the mangrove forests using satellite imagery at medium to high resolution, EEC Research Project No. 15017-1999-05 FIED ISP FR. Joint Research Center, Ispra

Alvarez-León R, Garcia-Hansen I (2003) Biodiversity associated with mangroves in Colombia. ISME/GLOMIS 3(1):1–2

Armitage D (2002) Socio-institutional dynamics and the political ecology of mangrove forest conservation in Central Sulawesi, Indonesia. Glob Environ Chang 12(3):203–217. https://doi.org/10.1016/S0959-3780(02)00023-7

Arriaga L, Montaño M, Vásconez J (1999) Integrated management perspectives of the Bahia de Caráquez Zone and Chone River Estuary, Ecuador. Ocean Coast Manag 42(2–4):229–241. https://doi.org/10.1016/s0964-5691(98)00055-6

Bakhuizen van den Brink R (1921) Revisio generis Avicenniae. Bull Jard Bot Buitenz 3:199–226

Batagoda BMS (2003) The economic valuation of alternative uses of mangrove forests in Sri Lanka. UNEP/GPA, The Hague

Blaber SJM (2007) Mangroves and fishes: issues of diversity, dependence, and dogma. Bull Mar Sci 80(3):457–472

Blanchard J, Prado G (1995) Natural regeneration of rhizophora mangle in strip clearcuts in Northwest Ecuador. Biotropica 27(2):160–167

Blanco JF (2018) Mangroves species richness: feedback. Research Gate, Berlin. https://doi.org/10.13140/RG.2.2.22958.15680

Bodero A (1993) Mangrove ecosystems of Ecuador. In: Lacerda LD (ed) Conservation and sustainable utilization of mangrove forests in Latin America and Africa regions. ISME, Okinawa, pp 55–74

Borda CA, Cruz R (2004) Pesca artesanal de bivalvos (Anadara tuberculosa y A. similis) y su relación con eventos ambientales. Pacífico colombiano. Rev Investig Mar 25(3):197–208

Boyd CE, Clay JW (1998) Shrimp aquaculture and the environment. Sci Am 278(June):58–65

Bunting P, Rosenqvist A, Lucas R, Rebelo L-M, Hilarides L, Thomas N, Hardy A, Itoh T, Shimada M, Finlayson C (2018) The global mangrove watch—a new 2010 global baseline of mangrove extent. Remote Sens 10(10):1669

Butler RA (2018) Mongabay: calculating deforestation figures for the Amazon. Mongabay. https://data.mongabay.com/general_tables.htm, https://data.mongabay.com/deforestation.htm, https://rainforests.mongabay.com/amazon/deforestation_calculations.html. Accessed 26 Jan 2017

Cárdenas H, Fuentes Á, González F, Quesada S, Gil J, Gallo J (2018) Conectividad genética de poblaciones naturales de la piangua (Anadara tuberculosa y Anadara similis) en la costa Pacífica colombiana estimada a partir de marcadores moleculares microsatélites [recurso electrónico]

Chong VC (2007) Mangroves-fisheries linkages in the Malaysian perspective. Bull Mar Sci 80(3):755–772

CLIRSEN (1987) Estudio Multitemporal de los Manglares, Camaroneras y Areas Salinas de la Costa Ecuatoriana. vol 1. Centro De Levantamientos Integrados De Recursos Naturales Por Sensores Remotos, Quito

CLIRSEN (2007) Actualizacion Del Estudio Multitemporal de Manglares, Camaroneras Y Areas Salinas En Las Costa Continental Ecuatoriana Al Ano 2006. vol 1. Centro De Levantamientos Integrados De Recursos Naturales Por Sensores Remotos, Quito

Collins S (2010) Mangrove destruction and shrimp aquaculture in ecuador: a focus on property right enforcement. University of Ottawa, Ottawa

Conchedda G, Lambin EF, Mayaux P (2011) Between land and sea: livelihoods and environmental changes in mangrove ecosystems of senegal. Ann Assoc Am Geogr 101(6):1259–1284. https://doi.org/10.1080/00045608.2011.579534

Costanza R, d'Arge R, de Groot R, Farberk S, Grasso M, Hannon B, Limburg K, Naeem S, O'Neill RV, Paruelo J, Raskin RG, Sutton P, van den Belt M (1997) The value of the world's ecosystem services and natural capital. Nature 387:253–260. https://doi.org/10.1038/387253a0

Costanza R, de Groot R, Sutton P, van der Ploeg S, Anderson SJ, Kubiszewski I, Farber S, Turner RK (2014) Changes in the global value of ecosystem services. Glob Environ Chang 26:152–158. https://doi.org/10.1016/j.gloenvcha.2014.04.002

Doat J (1977) The calorific value of tropical woods. Bois et Forets des Tropiques 172:33–55

Dodson CH, Gentry AH (1991) Biological extinction in Western Ecuador. Ann Mo Bot Gard 78(2):273–295

Donato DC, Kauffman JB, Murdiyarso D, Kurnianto S, Stidham M, Kanninen M (2011) Mangroves among the most carbon-rich forests in the tropics. Nat Geosci 4(5):293–297

Duke N (2018) On the difference between *R. Mangle* and *R. Samoensis*. Personal Communication 18th December 2018

Ebarvia M, Corazón M (1999) Total economic valuation: coastal and marine resources in the straits of Malacca. MPP-EAS Technical Report No. 24. PEMSEA Technical Report. GEF/UNDP/IMO Regional Programme for the Prevention and Management of Marine Pollution in the East Asian Seas (MPP-EAS)/Partnerships in Environmental Management for the Seas of East Asia (PEMSEA), Quezon City, Phillipines

Ellison JC, Duke N (2015) Rhizophora samoensis. The IUCN red list of threatened species version 2018-2. IUCN. Accessed 12 Dec 2018

Ellison AM, Farnsworth EJ (1996) Anthropogenic disturbance of caribbean mangrove ecosystems: past impacts, present trends, and future predictions. Biotropica 28(4):549–565. https://doi.org/10.2307/2389096

Ellison AM, Farnsworth EJ, Merkt RE (1999) Origins of mangrove ecosystems and the mangrove biodiversity anomaly. Glob Ecol Biogeogr 8(2):95–115

Ellison A, Farnsworth E, Moore G (2015) Rhizophora mangle. The IUCN red list of threatened species version 2018-2. IUCN. Accessed 12 Dec 2018

Fisher P, Spalding M (1993) Protected areas with mangrove habitat. vol draft report. World Conservation Centre, Cambridge

Flater D (2018) Tide table: Guayaquil, Ecuador. http://tides.mobilegeographics.com/calendar/year/2324.html. Accessed 24 Nov 2018

Frias-Torres S, Barroso P, Eklund A-M, Schull J, Serafy JE (2007) Activity patterns of three juvenile goliath grouper, Epinephelus Itajara, in a mangrove nursery. Bull Mar Sci 80:587–594

Giri C, Ochieng E, Tieszen LL, Zhu Z, Singh A, Loveland T, Masek J, Duke N (2011) Status and distribution of mangrove forests of the world using earth observation satellite data. Glob Ecol Biogeogr 20(1):154–159. https://doi.org/10.1111/j.1466-8238.2010.00584.x

Gore R (1977) The tree nobody liked. Natl Geogr 155:669–689

Graindorge D, Calahorrano A, Charvis P, Collot JY, Bethoux N (2004) Deep structures of the Ecuador convergent margin and the Carnegie Ridge, possible consequence on great earthquakes recurrence interval. Geophys Res Lett 31(4):L04603

Granek EF, Frasier K (2007) The impacts of red mangrove (Rhizophora Mangle) deforestation on zooplankton communities in Bocas Del Toro, Panama. Bull Mar Sci 80(3):905–914

Groombridge B (1992) Global biodiversity: status of the earth's living resources. WCMC, London

Guevara JM, Granda V (2009) El manglar es vida. Ministerio de Cultura, Quito

Hamilton SE (2013) Assessing the role of commercial aquaculture in displacing mangrove forest. Bull Mar Sci 89(2):585–601. https://doi.org/10.5343/bms.2012.1069

Hamilton SE, Casey D (2016) Creation of a high spatio-temporal resolution global database of continuous mangrove forest cover for the 21st century (CGMFC-21). Glob Ecol Biogeogr 25(6):729–738. https://doi.org/10.1111/geb.12449

Hamilton SE, Collins S (2013) Las Respuestas a Los Medios De Subsistencia Deforestación De Los Manglares en Las Provincias Del Norte De Ecuador. Bosque 34(2):143–153. https://doi.org/10.4067/S0717-92002013000200003

Hamilton SE, Friess DA (2018) Global carbon stocks and potential emissions due to mangrove deforestation from 2000 to 2012. Nat Clim Chang 8(3):240–244. https://doi.org/10.1038/s41558-018-0090-4

Hamilton SE, Lovette J (2015) Ecuador's mangrove forest carbon stocks: a spatiotemporal analysis of living carbon holdings and their depletion since the advent of commercial aquaculture. PLoS One 10(3):e0118880. https://doi.org/10.1371/journal.pone.0118880

Hamilton SE, Stankwitz C (2012) Examining the relationship between international aid and mangrove deforestation in coastal Ecuador from 1970 to 2006. J Land Use Sci 7(2):177–202. https://doi.org/10.1080/1747423x.2010.550694

Hamilton SE, Lovette JP, Borbor-Cordova MJ, Millones M (2017) The carbon holdings of Northern Ecuador's mangrove forests. Ann Am Assoc Geogr 107(1):54–71. https://doi.org/10.1080/24694452.2016.1226160

Hamilton SE, Castellanos-Galindo GA, Millones-Mayer M, Chen M (2018) Remote sensing of mangrove forests: current techniques and existing databases. In: Makowski C, Finkl CW (eds) Threats to mangrove forests: hazards, vulnerability, and management, vol 25. Springer International Publishing, Cham, pp 497–520. https://doi.org/10.1007/978-3-319-73016-5_22

Hansen MC, Potapov PV, Moore R, Hancher M, Turubanova SA, Tyukavina A, Thau D, Stehman SV, Goetz SJ, Loveland TR, Kommareddy A, Egorov A, Chini L, Justice CO, Townshend JRG (2013) High-resolution global maps of 21st-century forest cover change. Science 342(6160):850–853. https://doi.org/10.1126/science.1244693

Harcourt CS, Sayer JA (1996) The conservation atlas of tropical forests: the Americas. MacMillan Reference Books, New York

He Z, Zhang Z, Guo W, Zhang Y, Zhou R, Shi S (2015) De novo assembly of coding sequences of the mangrove palm (Nypa fruticans) using RNA-Seq and discovery of whole-genome duplications in the ancestor of palms. PLoS One 10(12):e0145385

Hogarth PJ (1999) The biology of mangroves. Biology of habitats. Oxford University Press, New York

Hogarth PJ (2007) The biology of mangroves and seagrasses. Biology of habitats. Oxford University Press, Oxford

Hogarth PJ (2015) The biology of mangroves and seagrasses. Oxford University Press, Oxford

Hutchison J, Spalding M, zu Ermgassen P (2014) The role of mangroves in fisheries enhancement. The Nature Conservancy and Wetlands International, Cambridge, p 54

IFAS (2009) Coastal wetland plants. University of Florida. http://fmel.ifas.ufl.edu/habitat/cono.shtml. Accessed 30 Jan 2009

IMF (2017) World economic outlook database. World economic and financial surveys. International Monetary Fund, Washington, DC

ITTO, ISME (1993) Mangrove ecosystems: technical reports. International Society for Mangrove Ecosystems (ISME), International Tropical Timber Organization (ITTO), Japan International Association for Mangroves (JIAM), Okinawa

IUCN SSC (2010) IUCN red list categories and criteria: version 3.1

IUCN, UNEP (2016) The World Database on Protected Areas (WDPA). www.protectedplanet.net. Accessed 01 Dec 2018

Kathiresan K, Qasim SZ (2005) Biodiversity of mangrove ecosystems. Hindustan Pub. Corp, New Delhi

Kelvin K, Lim P, Murphy DH, Morgany T, Sivasothi N, Peter K, Ng L, Soong BC, Hugh TW, Tan KS, Tan T, Tan K (2001) A guide to mangroves of Singapore. Raffles Museum of Biodiversity Research, The National University of Singapore & The Singapore Science Centre. Accessed 01 Apr 2008

Koenig CC, Coleman FC, Eklund A-M, Schull J, Ueland J (2007) Mangroves as essential nursery habitat for goliath grouper (Epinephelus itajara). Bull Mar Sci 80(3):567–586

Labastida E (1995) Diagnóstico Económico de las Actividades Relacionadas con la Zona de Manglar con énfasis en las Unidades de Producción Camaronera. In: Inefan E (ed) Estudio de las Alternativas de Manejo del Área Comprendida entre los Ríos Cayapas y Mataje, Provincia de Esmeraldas, vol 1. EcoCiencia & INEFAN, Quito, p Appendix 9

López-Angarita J, Roberts CM, Tilley A, Hawkins JP, Cooke RG (2016) Mangroves and people: lessons from a history of use and abuse in four Latin American countries. For Ecol Manag 368:151–162

Lugendo BR, Nagelkerken I, Kruitwagen G, van der Velde G, Mgaya YD (2007) Relative importance of mangroves as feeding habitats for fishes: a comparison between mangrove habitats with different settings. Bull Mar Sci 80(3):497–512

Lugo AE, Snedaker SC (1974) The ecology of mangroves. Annu Rev Ecol Syst 5:39–64

Macnae W (1968) A general account of the fauna and flora of mangrove swamps and forests in the Indo-West Pacific region. Adv Mar Biol 6:73–270

Macnae W, Kalk M (1962) The ecology of the mangrove Swamps at Inhaca Island, Mocambique. J Ecol 50(1):19–34

Madsen JE, Mix R, Balslev H (2001) Flora of Puna Island: Plant resources on a neotropical Island. Aarhus University Press, Langelandsgade

Mejía-Rentería JC, Castellanos-Galindo GA, Cantera-Kintz JR, Hamilton SE (2018) A comparison of Colombian Pacific mangrove extent estimations: implications for the conservation of a unique Neotropical tidal forest. Estuar Coast Shelf Sci 212:233–240. https://doi.org/10.1016/j.ecss.2018.07.020

Moreno J, Mora E (2009) La pesquería artesanal del recurso concha Andara tuberculosa y A. similis en la costa ecuatoriana durante el 2004

Nagelkerken I (2007) Are non-estuarine mangroves connected to coral reefs through fish migration? Bull Mar Sci 80(3):595–607

Naylor RL, Goldburg RJ, Mooney H, Beveridge M, Clay J, Folke C, Kautsky N, Lubchenco J, Primavera J, Williams M (1998) Nature's subsidies to shrimp and salmon farming. Science 282(5390):883–884

O'Dea A, Lessios HA, Coates AG, Eytan RI, Restrepo-Moreno SA, Cione AL, Collins LS, de Queiroz A, Farris DW, Norris RD, Stallard RF, Woodburne MO, Aguilera O, Aubry M-P, Berggren WA, Budd AF, Cozzuol MA, Coppard SE, Duque-Caro H, Finnegan S, Gasparini GM, Grossman EL, Johnson KG, Keigwin LD, Knowlton N, Leigh EG, Leonard-Pingel JS, Marko PB, Pyenson ND, Rachello-Dolmen PG, Soibelzon E, Soibelzon L, Todd JA, Vermeij GJ, Jackson JBC (2016) Formation of the Isthmus of Panama. Sci Adv 2(8):e1600883. https://doi.org/10.1126/sciadv.1600883

Ocampo-Thomason P (2006) Mangroves, people and cockles: impacts of the shrimp-farming industry on mangrove communities in Esmeraldas Province, Ecuador. In: Hoanh CT, Tuong TP, Gowing JW, Hardy B (eds) Environment and livelihoods in tropical coastal zones: managing agriculture-fishery-aquaculture conflicts (Comprehensive assessment of water management in agriculture series), vol 2. Oxford University Press, London, pp 140–153

Odum WE, Heald EJ (1972) Trophic analyses of an estuarine mangrove community. Bull Mar Sci 22(3):671–738

Ong JE (2002) The hidden costs of mangrove services: use of mangroves for shrimp aquaculture. Paper presented at the International Science Roundtable for the Media, Bali, Indonesia, June 04, 2002

Palomares MLD, Pauly D (2018) Sealifebase: Anadara tuberculosa v10.2018. Vancover.

Parks PJ, Bonifaz M (1994) Nonsustainable use of renewable resources: mangrove deforestation and mariculture in Ecuador. Mar Resour Econ 9(1):1–18

Plaziat J-C, Cavagnetto C, Koeniguer J-C, Baltzer F (2001) History and biogeography of the mangrove ecosystem, based on a critical reassessment of the paleontological record. Wetl Ecol Manag 9(3):161–180

Rosati I, Prosperi P, Latham J, Kainuma M (2008) World atlas of mangroves. In: Sessa R (ed) Terrestrial observations of our planet, GTOS 50, vol 1. Food and Agricultural Organization of the United Nations, Rome, pp 30–31

Saenger P, Hegerl EJ, Davie JDS (1983) Global Status of mangrove ecosystems, Commission on Ecology Papers No. 3, vol 3. World Conservation Union (IUCN), Gland

SeaLifeBase: Anadara tuberculosa V10.2018 (2018) https://www.sealifebase.ca/summary/Anadara-tuberculosa.html https://www.sealifebase.ca/. Accessed 10 2018

Shervette VR, Aguirre WE, Blacio E, Cevallos R, Gonzalez M, Pozo F, Gelwick F (2007) Fish communities of a disturbed mangrove wetland and an adjacent Tidal River in Palmar, Ecuador. Estuar Coast Shelf Sci 72(1–2):115–128

Siikamäki J, Sanchirico JN, Jardine SL (2012) Global economic potential for reducing carbon dioxide emissions from mangrove loss. Proc Natl Acad Sci Early Ed 109:14369–14374. https://doi.org/10.1073/pnas.1200519109

Snedaker SC (1986) Traditional uses of south american mangrove resources and the socio-economic effect of ecosystem changes. In: Kunstadter P, Bird ECF, Sabhasri S (eds) Workshop on man in the mangroves, vol 1. United Nations University, Tokyo, pp 102–112

Spalding M, Blasco F, Field C (eds) (1997) World mangrove atlas. International Society for Mangrove Ecosystems, Okinawa

Spalding M, Kainuma M, Collins L (2010) World atlas of mangroves. Earthscan, London

Spalding M, McIvor A, Tonneijck F, Tol S, Eijk PV (2014) Mangroves for coastal defence. Wetlands Internat, Wageningen

Taillardat P, Friess DA, Lupascu M (2018) Mangrove blue carbon strategies for climate change mitigation are most effective at the national scale. Biol Lett 14(10):20180251

Tomascik T (1997) The ecology of the Indonesian seas. Oxford University Press, Oxford

Tomlinson PB (1986) The botany of mangroves. Cambridge University Press, Cambridge/Cambridgeshire

Twilley RR, Cárdenas W, Rivera-Monroy VH, Espinoza J, Suescum R, Armijos MM, Solórzano L (2001) The gulf of guayaquil and the Guayas River Estuary, Ecuador. In: Seeliger U, Kjerfve B (eds) Coastal marine ecosystems of Latin America. Springer Berlin Heidelberg, Berlin/Heidelberg, pp 245–263. https://doi.org/10.1007/978-3-662-04482-7_18

UN FAO (1981) Tropical forest resources assessment project: forest resources of tropical asia. FAO, Rome

UN FAO (2004) Mangrove forest management guidelines, FAO Forestry Paper 117, vol 117. FAO, Rome. ISBN 92-5-103445-1

UN FAO (2007) The world's mangroves 1980–2005, FAO Forestry Paper, vol 153. FAO, Rome

UN FAO Fisheries and Aquaculture Department (2017) FishStat Plus – Universal software for fishery statistical time series. FAO, Rome

UN FAO Fisheries and Aquaculture Department (2018) The State of world fisheries and aquaculture 2018—Meeting the sustainable development goals. The state of world fisheries and aquaculture. UN FAO Fisheries and Aquaculture Department, Rome

Valiela I, Bowen JL, York JK (2001) Mangrove forests: one of the world's threatened major tropical environments. Bioscience 51(10):807–815. https://doi.org/10.1641/0006-3568(2001)051[0807,MFOOTW]2.0.CO:2

Veach K (1996) Gender, resource use, conservation attitudes and local participation in mangrove fishing villages in Northern Esmeraldas Province, Ecuador. University of Florida, Gainesville

Wang L, Mu M, Li X, Lin P, Wang W (2011) Differentiation between true mangroves and mangrove associates based on leaf traits and salt contents. J Plant Ecol 4(4):292–301. https://doi.org/10.1093/jpe/rtq008

Warne K (2007) Mangroves: forests of the tide. Natl Geogr 211:88–117

Yamanaka MD, Ogino S-Y, Wu P-M, Jun-Ichi H, Mori S, Matsumoto J, Syamsudin F (2018) Maritime continent coastlines controlling earth's climate. Prog Earth Planet Sci 5(1):21. https://doi.org/10.1186/s40645-018-0174-9

Yáñez-Espinosa L, Flores J (2011) A review of sea-level rise effect on mangrove forest species: anatomical and morphological modifications. In: Casalegno S (ed) Global warming impacts-case studies on the economy, human health, and on urban and natural environments. InTech Europe, Rijeka, Croatia. ISBN: 978-953-307-785-7

Zhang Z, He Z, Xu S, Li X, Guo W, Yang Y, Zhong C, Zhou R, Shi S (2016) Transcriptome analyses provide insights into the phylogeny and adaptive evolution of the mangrove fern genus Acrostichum. Sci Rep 6:35634

Zorn J (1796) Afbeeldingen der artseny-gewassen met derzelver nederduitsche en latynsche beschryvingen. In: Oskamp L, Houttuyn M, Krauss JC (eds) Images of the physician's crops with their traditional and Latin names

Chapter 2
Shrimp Farming

Abstract This chapter examines commercial shrimp farming aquaculture systems in Ecuador and beyond. Section 2.1 identifies the remarkable growth in aquaculture and shrimp aquaculture production since the advent of commercial shrimp farming in about 1970 and expands on the role of international finance in driving this expansion. Section 2.2, describes the shrimp aquaculture process in Ecuador from the initial hatching to the final harvesting of shrimp. Section 2.3, describes the entire life cycle of *P. vannamei* in both the wild and in an aquaculture setting and Sect. 2.4 expands on this by discussing the importance of *P. vannamei* to the nearshore environment of Ecuador. Section 2.5 examines the human dimension of shrimp farm aquaculture in Ecuador with a focus on tractional fishing communities and Sect. 2.6 examines the legal framework of shrimp aquaculture, particularly those now resining in former mangrove forests.

Keywords Shrimp farming · Shrimp aquaculture · *P. vannamei* · Whiteleg shrimp

2.1 Ten Thousand Years of Development in Fifty Years: The Growth of Commercial Aquaculture

Traditional coastal communities have reared shrimp in humanmade enclosures for centuries. This form of traditional shrimp aquaculture focuses on local cultivation of the native shrimp species within the mangrove forest and with limited mangrove forest clearing. The general method applied is to trap wild shrimp inside small earthen mounds within the mangrove forest and harvest them as required. Preexisting nutrients provide the conditions for shrimp growth. This form of small-scale shrimp farming is for local consumption utilizing the local environmental conditions. These aquaculture practices continue currently in isolated coastal communities in parts of Bangladesh, Malaysia, Ecuador, and Indonesia. Such methods of aquaculture are more akin to traditional fishing practices and do not relate to the modern concept of commercial shrimp aquaculture that has taken hold in the last 40-years.

S. E. Hamilton, *Mangroves and Aquaculture*, Coastal Research Library 33,
https://doi.org/10.1007/978-3-030-22240-6_2

By any unit of measure, and when viewed from either side of the scientific debate concerning the environmental merits of shrimp aquaculture (Boyd and Clay 1998), the growth of commercial aquaculture since 1970 has been remarkable and rapid. In 1970, less than 4% of seafood consumed worldwide was reared in a farmed environment (UN FAO Fisheries and Aquaculture Department 2018c). By 2015, aquaculture had surpassed wild catch as the dominant source of human seafood consumption, surpassing the 50% mark; and by 2016, it has risen to provide 53% of all seafood consumed by humans (UN FAO Fisheries and Aquaculture Department 2018c).

Aquaculture will likely continue to be an expanding production system, with growth rates continuing to exceed 10% annually (Diana 2009). Within emerging economies, aquaculture growth has often exceeded 10% annually since 1970 (UN FAO Fisheries and Aquaculture Department 2006). Globally, aquaculture grew at 11.3% during the 1980s and 10% during the 1990s, but aquaculture growth has slowed to 5.8% since 2000 (Fig. 2.1) (UN FAO Fisheries and Aquaculture Department 2018c). As a comparison, farmed meat production grew by an average of 2.8% for much of this period (UN FAO Fisheries and Aquaculture Department 2006; United Nations Committee on World Food Security 2003). In 1984, global aquaculture was still a nascent industry, but by 2016, it was annually producing more than 110 million tonnes of seafood valued at almost $250 billion (Fig. 2.1).

Within 50 years of being a nominal and highly-specialized activity producing almost no seafood, commercial aquaculture production has surpassed wild catch as the primary source of seafood in human diets. Although difficult to compare, it is worth noting that the conversion from wild terrestrial animal stock to farmed terrestrial stock took many tens of thousands of years.

Crustacean aquaculture farming is now one of the highest dollar value components of aquaculture globally and is worth over $57 billion annually. This value equates to 23% of the entire global aquaculture market. (UN FAO Fisheries and

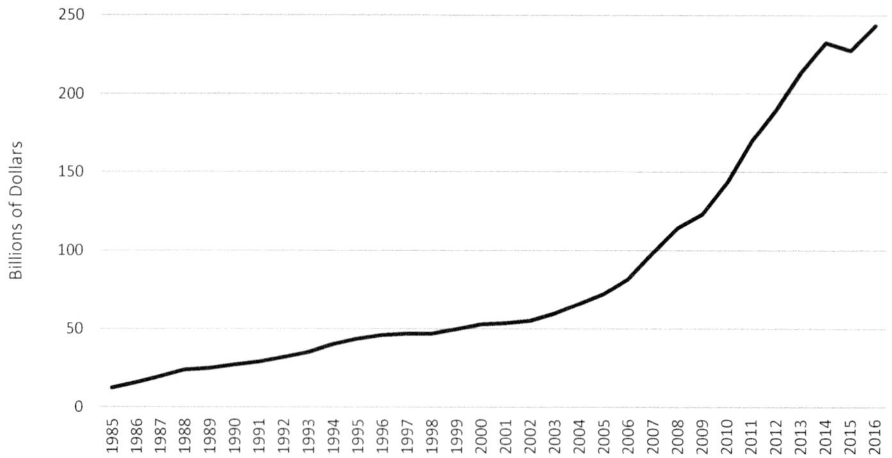

Fig. 2.1 Global aquaculture output in USD from 1982 to 2016. (UN FAO Fisheries and Aquaculture Department 2017)

Aquaculture Department 2017, 2018c). One single shrimp species, Whiteleg shrimp (*Penaeus vannamei*), constitutes 53% of the entire global crustacean aquaculture market (UN FAO Fisheries and Aquaculture Department 2017, 2018c), which equates to over 12% of all commercial aquaculture value across all species and all regions globally. As of now, almost 8 billion tonnes of *P. vannamei* are produced in the shrimp farms of the world (UN FAO Fisheries and Aquaculture Department 2017).

Shrimp farming as a viable large-scale commercial aquaculture process began in both Crystal River, Florida and Panama City, Florida (Rosenberry 2008; Cheshire 2005) in the mid-1960s. It was first documented in western geographic literature by National Geographic in 1965 (Idyll 1965). Both the Crystal River and Panama City enterprises utilized United States capital and Japanese scientific expertise. A group of former DuPont employees began the Panama City facility and Purina directly funded the Crystal River facility. It was soon realized that the lack of postlarvae in local areas and the water temperature of the Gulf Coast of Florida made Florida inappropriate for successful commercial shrimp production (Rosenberry 2008; Cheshire 2005). After achieving the first successful spawning in 1973 (UN FAO 2014), the Crystal River operation relocated to Panama City, Panama soon after (Rosenberry 2008) but ceased operations by the mid-1970s, after this closure many US-based shrimp aquaculture experts relocated to Ecuador continuing the quest for farmed shrimp (Cheshire 2005).

In 1974, staff from the Panama City hatchery, operated by Marifarms, visited Esmeraldas, Ecuador to obtain *P. vannamei* broodstock, and noted that environmental conditions in Ecuador were suitable for year-round shrimp farming due to the water temperature and natural richness of feed in tidal waters. After the failure of the Panama City shrimp farm, Marifarm employees returned to Ecuador and collaborated with Ecuadorian producers to establish a hatchery in Manta (Cheshire 2005). By 1970, mass commercial farming of shrimp was underway in Ecuador. Indeed, Ecuador was the first nation globally to achieve true large-scale export-oriented shrimp farming.

Aquaculture production in Ecuador mirrors the global trend. In 1970, *P. vannamei* production in Ecuador was at approximately 50 tonnes per year (Fig. 2.2). By 2016, it had grown to 422,000 tonnes, (Fig. 2.2), with a value of $2.2 billion annually (UN FAO Fisheries and Aquaculture Department 2017). The entire Ecuadorian economy was worth approximately $97 billion in 2016 (World Bank 2017). Therefore, shrimp aquaculture is approximately 2.3% of the entire Ecuadorian economy. Ecuadorian shrimp exports were valued at $2.6 billion in 2016 (World Bank 2017); this is higher than the first-sale value listed above due to added value through processing. Thus, farmed *P. vannamei* alone constituted 15.3% of all of Ecuador's exports in 2016 (World Bank 2017). Asia is the current primary market for Ecuadorian farmed shrimp (Rodríguez 2018), with 64% of exports going there in 2017. In the early years of the shrimp industry, the primary market for Ecuadorian shrimp was the United States (Snedaker 1986). This export-oriented sector of the economy is visible across the landscape of coastal Ecuador, which is now dotted with shrimp growout ponds.

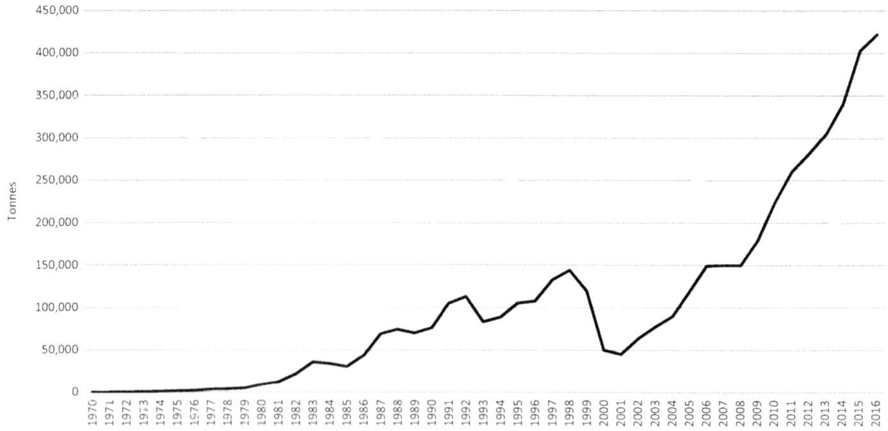

Fig. 2.2 Ecuadorian shrimp aquaculture output from 1970 to 2016. (UN FAO Fisheries and Aquaculture Department 2017)

2.1.1 Shrimp Farm Financing

The initial creation of shrimp growout ponds, shrimp hatcheries, and shrimp nurseries in more intensive systems require substantial capital investment (Asian Development Bank 1991; Rajitha et al. 2007; Flaherty and Karnjanakesorn 1995; Tobey et al. 1998). Even when operational, the shrimp farming system is a capital-intensive operation as opposed to labor-intensive. For example, in semi-intensive shrimp farming systems such as Ecuador, it is estimated that labor only constitutes 4% of the total operational cost, and feed constitutes the highest cost of an operating farm (Shang 1992). The source of this initial investment appears to have primarily been international development assistance.

The financing structure of the first shrimp aquaculture systems is difficult to determine, as much of the data is lost to history. Money flows were often bilateral and were processed before open records regarding international aid were commonplace. Additionally, as the awareness of the severe environmental impacts that can accompany aquaculture, and shrimp farm development, has increased, donors may be reluctant to tag their aid as directly supporting the construction of shrimp farms. Indeed, searches of aid databases show much of the money that flowed and still flows in aquaculture is defined as infrastructure development aid, social aid, rural development aid, or fisheries aid. Researchers seeking to identify aquaculture funding are required to find the implementation reports or evaluations to discover that much of the money went into shrimp pond construction or support of the shrimp farm industry.

During the 1970s and 1980s, the initial development and investment in shrimp farming across the developing world was often driven by investment banks, including the World Bank International Finance Corporation, the Asian Development Bank, the International Bank for Reconstruction and Development, the Japanese

Development Bank, the European Investment Bank, and the African Development Bank along with International Monetary Fund (IMF) support (Skladany and Harris 1995; Public Citizen's Food Program 2005; Hamilton and Stankwitz 2012). The largest contributions from bilateral aid agencies came from the United States Agency for International Development (USAID), the Overseas Economic Cooperation Fund, the Japanese International Cooperation Agency, the Canadian International Development Agency, and the Norwegian Agency for Development Cooperation (Public Citizen's Food Program 2005; Hamilton and Stankwitz 2012). Substantial technical support also came from the United Nations (UN). Other documentation of international aid funding aquaculture is captured in the peer-reviewed academic literature. For example, in the 6 years from 1997 to 1983, aquaculture was supported with financing of $476 million from development agencies alone (Nash 1987), and between 1988 and 1995, almost $1 billion was provided by aid agencies (Shehadeh and Orzeszk 1997). This financing is likely merely the tip of funding as financing from outside of the Organisation for Economic Co-operation and Development's (OECD) Development Assistance Committee (DAC) is not included, and neither is foreign direct investment (FDI), nor funding from private donors, or internally funding programs.

Public Citizen's Food Program (2005) provide a referenced series of foreign aid commitments of $1.8 billion into Latin America starting in 1980 and ending in 2000. Between 1998 and 2006, the International Finance Corporation (IFC) division of the World Bank made loans of $71.4 million to support and expand shrimp farming across Latin America, Asia, and Africa, and in 2006, they were preparing to finance a 20,000 ha series of shrimp farms in Indonesia (Wood and Mayer 2007). Interestingly, to minimize the implications of mangrove forest loss resulting from aquaculture, the IFC use the statistic that shrimp farming is only responsible for 10% of mangrove deforestation (Boyd and Clay 1998). This 10% figure refers to pre-industrial revolution mangrove cover and is often used out of context. When the mangrove forest cover data are processed for recent timeframes, the conversion of mangrove forest to shrimp farms is likely far higher (Hamilton 2015).

Development money flows into aquaculture originating from the developed world continue post-2000 to the present. Hamilton and Stankwitz (2012) state that in 2002, USAID gave $26 million to numerous projects to support livelihoods in Ecuador's northern border region, including projects that supported aquaculture (USAID 2003). Additionally, Hamilton and Stankwitz (2012) note that in a 2006 USAID-sponsored report, aquaculture was identified as an opportunity for economic growth in Iraq. This report stated that startup costs could be significantly lowered if aquaculture development occurred in a semi-intensive manner near the intersection of the Tigris and Euphrates Rivers (Louis Berger Group 2006). USAID is not alone in supporting aquaculture. Norway, Sweden, and Germany have historically supported aquaculture projects (Wood and Mayer 2007), and as of 2019, the United Kingdom national aid agency supports Msingi in Africa. Msingi is committed to supporting and financing aquaculture development within Lake Victoria (MSINGI 2010), which is a lake undergoing numerous environmental and biodiversity challenges.

Aquaculture across the developing world, including Ecuadorian shrimp farms, was most likely built on the foundation of international development financing through the numerous international, transnational, and national government aid agencies (Hamilton and Stankwitz 2012; Nash 1987; Shehadeh and Orzeszk 1997). Undoubtedly, local investment and private foreign direct investment also played a role, since, by the 1990s, liberal national investment policies and favorable state incentives had attracted transnational and national private investors (Wood and Mayer 2007). It appears these private monies flowed only after the initial rounds of development funding. This model of private funds growing a nasceant industry after development banks and aid organizations invested heavily in the formative development of the industry is considered a successful investment strategy and is the stated goal of many development agencies and aid agencies. Considering the current levels of global aquaculture output (Fig. 2.1) and the current levels of Ecuadorian shrimp farm output (Fig. 2.2), the growth of aquaculture and shrimp farming is a remarkable success story for development banks and international aid agencies.

2.2 Ecuadorian Shrimp Farm Construction and Technical Operation

In Ecuador, the shrimp farm industries growout ponds are generally located within the inter-tidal environment of numerous sheltered estuaries or in the mouths of coastal rivers. Shrimp growout ponds located near the mouths of estuaries are occasionally extensive systems that use the natural tidal flows of the premium site for free input feed, free input nutrients, free water exchange, and free aeration of the growout ponds. However, these extensive growout systems are the minority of pond aquaculture systems in Ecuador. Most ponds are located farther into the estuaries, within rivers, near the mouth of the river, on the edge of the estuaries, or on land close to the estuary. All these semi-intensive ponds are within an affordable pumping height and distance of the estuarine waters required to sustain them. Locating shrimp growout ponds higher than 8 m above the high tide region of an estuary would involve the use of expensive submersible pumping systems or multiple pumping stations and reservoirs, making the shrimp farming activity not economically competitive (Hamilton 2012).

A shrimp farm in Ecuador is analogous to a traditional terrestrial livestock farm. A terrestrial farm is made up of many fields; the shrimp farm equivalent to agricultural fields are the growout ponds. Just as animals are reared, fattened, and reside in fields of a farm, shrimp are reared, grown to edible size, and reside in the growout ponds of the farm. Other infrastructure likely exists on a terrestrial farm as well, such as housing, equipment storage, and a nursery or indoor location for juvenile animals. Some trees are often present as well, but most of a terrestrial farm consists of livestock fields. The same applies to shrimp farms. Other regions of a shrimp farm away from the ponds store equipment, house juvenile shrimp, often have a house or lookout tower and have feeder systems to move water between the estuary and the growout ponds. They may even have some mangrove trees, but the largest area of a shrimp farm is the growout ponds.

Shrimp farm growout ponds are generally grouped into three different classifications based on stocking densities and other characteristics; these three systems are generally known as extensive systems, semi-extensive systems, and semi-intensive systems (Twilley 1989; UN FAO 2014). Extensive growout pond stocking densities are low at 5000–30,000 juveniles per hectare, are shallow (Tobey et al. 1998; UN FAO 2014), and require few if any external inputs; with the sea water providing the required nutrients (Twilley 1989). Next, semi-extensive growout ponds have stocking densities between 25,000 and 200,000 juveniles (Tobey et al. 1998; UN FAO 2014) and use a combination of external inputs and natural inputs (Tobey et al. 1998; UN FAO 2014). Third, semi-intensive systems stock juveniles at over 200,000 (Tobey et al. 1998; UN FAO 2014) juveniles per hectare and require constant water exchange, regular high-protein feeding, fertilization, and numerous other external treatments (Twilley 1989). The semi-intensive growout ponds in Ecuador typically have between 100,000 and 300,000 juveniles per hectare (UN FAO 2014), uniform rectangular or square shapes, generally are larger than 5 ha, have water depths around 1.2 m, a pumping system, some aeration, formulated feed, average as many two to three crops per year, and have the high potential for disease.

Some confusion exists over the use of semi-extensive, semi-intensive, and intensive in the shrimp farming systems literature. In this manuscript, we use the term semi-intensive to mean growout ponds that are primarily driven by external feed, pumped water, medium to high stocking densities, minimal aeration, and have a high potential level for the disease. Semi-intensive is the term is used in Ecuador by Twilley (1989) and across Latin America by the UN FAO (2014). Other authors such as Tobey et al. (1998) use the term semi-extensive for these growout ponds.

As of the early1990s, extensive systems constituted approximately 60% of the growout ponds in Ecuador and semi-extensive for approximately 40% (Tobey et al. 1998). Since this time, almost all news farms are semi-intensive, as the primary areas for extensive farms have already been consumed and ponds moved further into the estuary, onto the banks of the estuary, or just outside the estuary, and none of these sites allow for extensive growout pond systems. Primarily though, the transition from extensive to semi-intensive was due to the lack of available estuarine space suitable for more extensive systems. Additionally, regulation forced newer growout ponds out of the prime estuarine space.

Semi-intensive growout ponds now dominate Ecuador's estuarine landscape with few extensive systems remaining. Figures 2.3, 2.4, and 2.5 depict the most common semi-intensive estuarine shrimp pond systems utilized in Ecuador. Figure 2.3 depicts a shrimp farm with about five or six growout ponds. The term shrimp farm describes the entire operation within the estuary, and the term shrimp pond refers typically to the growout ponds where the shrimp spend most of their life. Juvenile shrimp are reared in growout ponds until the shrimp reach maturity and market size and they are harvested. The growout period is typically around 3 months in Ecuador, with two or three crops per year harvested. The growout ponds

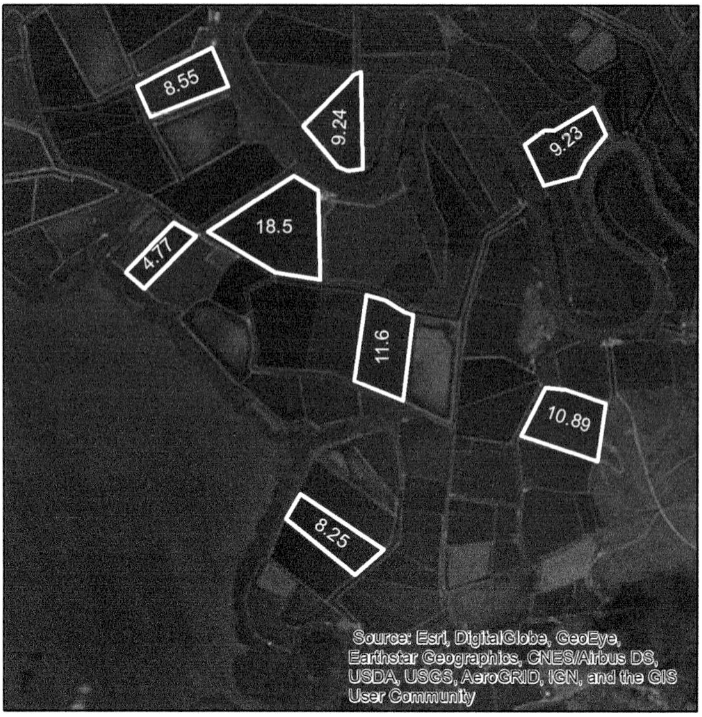

Fig. 2.3 Typical shrimp ponds in an Ecuadorian estuary. The interior of Cojimies estuary in Extreme Northern Manabí Province, 3 km Southeast of Chamanga, January 2016. The units are hectares. All the regular shapes are ponds, and a selection are outlined to demonstrate the general size

are divided from each other by earthen dikes up to 5 m high; one levee usually serves two ponds or one pond and the estuary edge. Each pond can vary in size from a few hectares to 50 ha. The ponds depicted in Fig. 2.3 are typically-sized growout ponds. Of the over 200 ponds have been visited by the author, none appear to be lined with plastic or similar. The most prestigious and productive farms appear to be those located near the mouth of the estuary but still sheltered from the waves, and the least productive and those that require the most inputs are in the interior of the estuary in the mouths of rivers that flow into the estuaries.

Figure 2.4 illustrates the water intake of a growout pond in an estuary. Growout ponds replace approximately 10% of their water daily, typically pumping this water from the estuary in which they are located on an almost continuous basis (Stram et al. 2005). For example, in a study of growout pond water exchange in a Colombian pond system, it was estimated that a typically-sized pond replaces 345 m^3–600 m^3 of water per day (Gautier et al. 2001). The two large pipes that are visible on the left of the image run from the wooden pumphouse on one end, into the estuary on the other end. The pipes typically extend a few feet below the surface of the water. Often one pipe is a backup and not operational. In Fig. 2.4, the left pipe is currently

Fig. 2.4 A typical Ecuadorian shrimp pond water intake system. (The photograph was taken near the village of Salinas on the northern side of Chone estuary, January 2008. Photographer Hamilton, S.E)

Fig. 2.5 A typical Ecuadorian shrimp pond discharge system. (The picture was taken near the village of Salinas on the northern side of Chone estuary, January 2008 by Hamilton, S.E)

disconnected and under repair. The wooden pumphouse in the image is located on the levee that separates the growout pond and the estuary. The shrimp pond is behind the wooden pump house. The pump, usually just one running on diesel or grid powered where mains electricity is present, is inside the pumphouse.

Figure 2.5 demonstrates the draining of water from a shrimp pond back into the estuary. Water intakes and outflows are usually located in different parts of the pond. The concrete and wooden dam structure allow for the return of water from the shrimp pond back into the estuary. The shrimp pond is located behind the outflow pictured. Numerous outflow systems exist, but a wooden and concrete system such as the one in Fig. 2.5 is most common. Wooden slats can be added or removed to limit or expand the concrete opening to raise or lower the level of the water in the growout pond. When draining the pond, a net or mesh is placed over the opening to capture any shrimp that may escape.

The lifespan of an estuarine shrimp pond is short. The length of shrimp pond productivity is dependent on the practices and location of the shrimp pond. In the literature, shrimp farm lifespan is given as 5–10 years (Naylor et al. 1998) in an intensive system or 7–15 years in a less intensive system, with initial high yields usually followed by a dramatic collapse (Paez-Osuna 2001). Bacterial contamination of the sediment at the base of the pond and viral disease are the primary reasons for shrimp pond abandonment and production collapses. There is general agreement that abandoned farms rarely regain their productivity after a collapse. Most authors also contend that abandoned farms are rarely rehabilitated back into mangrove forest ecosystems, nor are they converted to other agricultural uses (Naylor et al. 1998; Paez-Osuna 2001), although this fact is disputed (Diana 2009). Ecuadorian shrimp farms follow the general trends noted in the literature. Many abandoned farms along the coast are still visible on the landscape many years after abandonment despite the recent resurgence in Ecuadorian shrimp farm output, few if any ponds appear to return to aquaculture. Shrimp farmers do occasionally grow Tilapia in ponds once they have become too contaminated for shrimp and some farmers do report the use of abandoned ponds for Tilapia growout, but this practice appears uncommon. Figure 2.6 depicts numerous abandoned ponds on the fringes of the Chone Estuary. The dry ponds are in the foreground with grass growing on the former levees. Note that the former bottom of the ponds has little or no vegetation. The estuary is just visible in the background of the image and mangroves in the foreground.

2.3 The Entire Lifecycle of *P. vannamei* in an Ecuadorian Shrimp Farming System from Broodstock to Mature Shrimp

Semi-intensive shrimp farming, as commonly practiced in Ecuador, operates on an approximately 90-day cycle from the release of eggs by the female to the harvesting of adult shrimp. During this time, shrimp move from a hatchery to a nursery, and

Fig. 2.6 Abandoned shrimp ponds on the Fringe of the Chone estuary. (The picture was taken approximately 5 km east of Puerto Larrea on the southern side of Chone estuary, January 2008 by Hamilton, S.E)

finally into a growout pond. Shrimp pass through four distinct life stages during the first 10–12 days of their existence. The first stage of the shrimp lifecycle occurs in a hatchery which, in Ecuador, is typically located away from the growout ponds, with each significant shrimp farming region having one or two hatcheries. After mating in the hatchery, the female shrimp, known as broodstock, disperse the fertilized eggs into the water column where they hatch into nauplii (the first larval stage of the shrimp, Fig. 2.7) within 1 day (UN FAO 2014). It is not uncommon for a single female to produce hundreds of thousands of eggs in one hatching event, and many millions of eggs over a lifetime. No feed is required at the nauplius stage in the aquaculture or wild environments, as all necessary nutrients are contained within the nauplii yolk reserves. The nauplii stage lasts approximately 2 days (UN FAO 2014). Figure 2.7 depicts the nauplii stage of the shrimp lifecycle in an Ecuadorian hatchery, and the beaker likely contains more than 1000 nauplii.

After the nauplii stage of development, shrimp progress to zoea and remain in this state for 4–5 days when they enter mysis for 3 or 4 days (UN FAO 2014). During these two stages, algae provide all the necessary nutrients for growth. In aquaculture, the algae are provided in large swimming-pool-style larval rearing tanks, most often located at the hatchery house (Fig. 2.8). At this stage of their life cycle, the shrimp are in an indoor, or at least covered and controlled, environment (Fig. 2.8). The pictured tanks were filled with algae-rich water, giving them a green color, and were maintained at a temperature above 23 °C by covering them with thick plastic

Fig. 2.7 Nauplii stage of shrimp development in an Ecuadorian hatchery. (Town of Canoa, August 2009, taken by Hamilton S.E)

Fig. 2.8 Algae rich tanks housing zoea and mysis at an Ecuadorian hatchery. Canoa, August 2009. (Photograph by Hamilton, S.E)

sheeting and providing heat. Figure 2.8 depicts a cluster of about 12 tanks with each containing pre-shrimp at differing life-cycle stages.

After about 3–4 days, the mysis metamorphose into postlarvae and remain in the rearing tanks for a few more days (UN FAO 2014). Postlarvae are essentially juvenile shrimp, with a body form similar to an adult shrimp. At this time, the postlarvae are transported from the offsite hatchery to a shrimp nursery on the shrimp farm (Fig. 2.9). Sometimes the nursery stage is omitted, and the postlarvae remain in the

Fig. 2.9 Ecuadorian shrimp nursery on a shrimp farm near Chone, August 2009. (Photograph by Hamilton, S.E)

rearing tanks for 10–15 more days (UN FAO 2014) before being directly deposited to the estuarine growout ponds. Unlike the growout ponds, the nursery ponds are lined with plastic and are often in a closed and controlled environment. The controlled nursery environment, as opposed to the open environment of the growout ponds, is necessary because the postlarvae are more vulnerable to temperature and habitat changes, and other externalities at this nascent point of their life. The nursery ponds are typically a couple of meters wide by about 10 m long. The postlarvae are provided with high protein artificial feed while in the nursery to enable growth and to limit cannibalism.

After about another 14–21 days in the nursery, the postlarvae are released into a growout pond in the estuary (Figs. 2.3 and 2.10). The now juvenile shrimp are merely scooped up with small handheld nets and released. Juvenile shrimp spend approximately the next 3 months maturing into full-sized adult shrimp in these growout ponds. These ponds are in the intertidal bays and estuaries along the entire range of the tropical Pacific and Atlantic coastlines of Latin America and within all the study sites included in this manuscript along the Pacific Coastline of mainland Ecuador.

While in the growout ponds, *P. vannamei* shrimp are fed a diet of pellets primarily composed of fishmeal, krill, soybean, and shrimp byproducts such as shrimp heads. In some more technologically advanced ponds, the feed is provided via automatic feeders stationed within the pond, or more commonly, the feed is distributed by merely throwing it by hand into the pond from one of the surrounding levees. The average protein content of feed utilized in Ecuador is between 25% and 35%

Fig. 2.10 Panel left, collecting the Postlarvae from the nursery to relocate in the growout pond. Panel right, transferring Postlarvae from the Nursery to a Growout Pond. Chone, January 2008. (The Photographs are Taken by Hamilton, S.E)

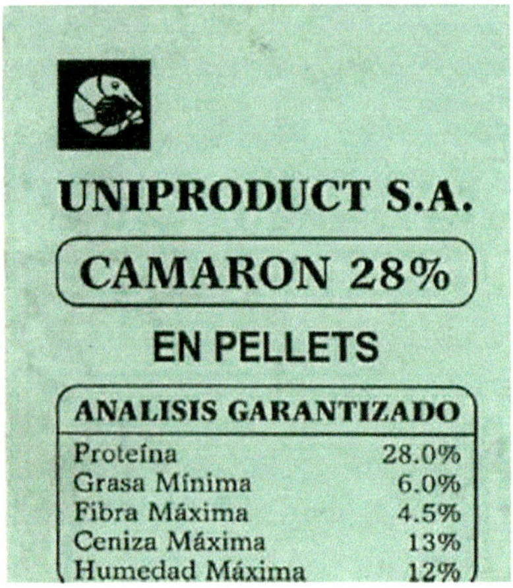

COMPANY NAME

SHRIMP 28%

PELLETS

ANALYSIS GUARANTEED

Protein	28.0%
Minimum Fat	6.0%
Maximum Fiber	4.5%
Maximum Ash (Phosphorous)	13%
Maximum Moisture	12%

Fig. 2.11 A shrimp feed tag taken from an Ecuadorian shrimp farm during feeding. (Translation provided by Hamilton, S.E)

(Fig. 2.11). Higher protein levels are generally applied earlier in the lifecycle when the shrimp are carnivorous in the wild, with protein feed levels dropping as the shrimp mature and in the wild would feast more on detritus in the estuary. Higher protein feed grades cost more money than lower protein feeds, thus, one of the benefits of rearing *P. vannamei* is that feed protein levels are as much as 20% lower than other shrimp species, resulting in reduced expenditure on feed (UN FAO 2014).

As of the mid-2000s, much of the protein in shrimp aquaculture feed (Fig. 2.11) comes from fishmeal generated by marine capture fisheries, with farmed shrimp consuming the most fishmeal globally compared to other species grown in aquaculture (Tacon and Metian 2008). Much of the input fishmeal for shrimp is generated from fish that are undesirable for human consumption with a feed conversion ratio

Fig. 2.12 Ocean fishmeal destined for use on an Ecuadorian shrimp farm. (The photograph was taken in a small fishing village north of Manta in January 2008 by Hamilton, S.E)

(FCR) of between 1.9: to 1.4 (Tacon and Metian 2008). Fishmeal is the preferred source of protein due to its high energy content, and it contains important minerals such as phosphorous, B vitamins, and required fatty acids (UN FAO Fisheries and Aquaculture Department 2018a). Fresh fishmeal has been shown to result in improved growth levels across all farmed species of shrimp (Ricque-Marie et al. 1998). Figure 2.12 depicts a full truck of fish destined for the shrimp processors of Manabí. It is estimated that during 2006, between 225,000 and 250,000 tonnes of wild fish were captured to sustain the Ecuadorian farmed shrimp industry with an FCR of 1.0–1.4 (Tacon and Metian 2008). This figure is likely to have doubled by 2016 as *P. vannamei* production has also doubled since 2006. Just as with *P. vannamei* hatcheries, feed-stores are often located away from the actual farms in a nearby urban center. As of 2018, a move towards terrestrial feed sources such as soy is underway in Ecuador.

Other inputs applied to growout ponds include herbicides to remove vegetation from the bottom of the ponds, pesticides to remove non-shrimp fauna, and lime to treat the sediment (Paez-Osuna 2001). These inputs are applied before and between shrimp growout cycles. Other chemical agents are added during the growout cycle to improve water quality (Naylor et al. 1998). Although, the European Union and the United States have banned the import of shrimp treated with antibiotics, and Ecuador supplies shrimp to both regions, shrimp antibiotic use appears widespread in Ecuadorian commercial shrimp aquaculture. Officially, no shrimp arrive in the European Union or the United States that is treated with antibiotics, yet on most farms observed, antibiotic use is a widespread and accepted practice. The economic hardship caused by a failed harvest due to disease makes antibiotic use a safeguard against negative economic returns. Each major shrimp-producing town has a store from which farmers purchase antibiotics for treatment in their feed and farms. The

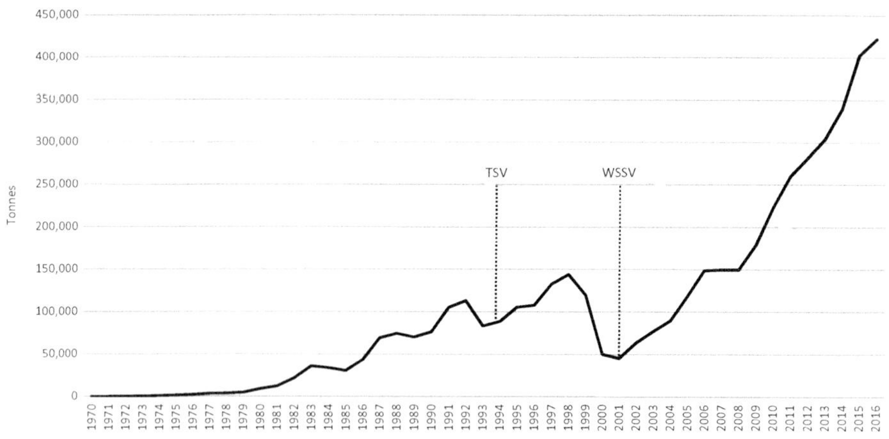

Fig. 2.13 Ecuador aquaculture output and disease outbreak

antibiotics mitigate potential losses from vibriosis, a bacterial disease that can occur throughout the farming process from the hatchery to the growout pond (UN FAO 2014).

In addition to bacterial diseases, viruses such as Taura syndrome virus (TSV) and white spot syndrome virus (WSSV) (McClennen 2004; Rodríguez et al. 2003) have periodically devastated the Ecuadorian farmed shrimp industry since the early 1990s. These viruses are not treated by antibiotics but are mitigated by the use of disinfectant and bleach of handling equipment as well as prevented by a host of other best practices (UN FAO 2014). Other viruses such as infectious hypodermal and haematopoietic necrosis (IHHNV) and baculoviral midgut gland necrosis (BMN) are also present in the shrimp farming industry (UN FAO 2014). Indeed, these viral, and to a lesser extent, bacterial disease outbreaks contribute to questions about the long-term benefit of shrimp aquaculture in general (Venizelos and Benetti 1996; Rodríguez et al. 2003; Hopkins et al. 1995). Such disease outbreaks not only shut down an entire estuary's commercial shrimp industry, but the catch of wild shrimp is also much reduced (Fig. 2.13).

2.4 The Importance of Ecuador's Coastal Estuaries to *P. Vannamei*

P. vannamei is the species under the decapod order, within the crustaceans' subphylum, that belongs to the Penaeidae family (Fig. 2.14). What makes penaeid shrimp almost unique among decapods is that they broadcast their fertilized eggs instead of carrying them until they hatch, and the larvae hatch out in the nauplius stage of their lifecycle (Wyban and Sweeney 1991). The species is catadromous; adults spawn in the Pacific Ocean close to shore, then postlarvae migrate inshore, burying

Fig. 2.14 *P. vannamei*. (Photo provided by the fish database of Taiwan)

P. vannamei

themselves in the soft mud of the shallow estuaries for a few months until they return to the nearshore ocean environments as adults (Wyban and Sweeney 1991).

P. vannamei are native to the tropical coastal waters of the Eastern Pacific (UN FAO Fisheries and Aquaculture Department 2018b). Their native range extends from the warm waters of the Gulf of California in Northern Mexico at the extreme northern end of the range, to northern Peru in the southern extreme of the range including the entire mainland coastal region of Ecuador and all the estuaries analyzed in this manuscript (UN FAO Fisheries and Aquaculture Department 2018b; Wyban and Sweeney 1991). Their primary habitats are areas where sea temperatures are above 20 °C year-round (UN FAO Fisheries and Aquaculture Department 2018b). Interestingly, this is almost the same isotherm within which mangrove forests thrive (Hogarth 2015; Tomlinson 1986). Indeed, the geographic range of *P. vannamei* (UN FAO Fisheries and Aquaculture Department 2018b; Wyban and Sweeney 1991) and the mangrove forest extent and locations along the tropical coastline of the American Pacific appear identical (Chapter 2, Figure 1.3). *P. vannamei* have been shipped all over the world for farm breeding because they are relatively easy to culture (Wyban and Sweeney 1991). Hence, *P. vannamei* aquaculture operations are now ubiquitous in Asia (Briggs et al. 2004).

Ecuador's Pacific coast provides an ideal environment for *P. vannamei* rearing due to its tropical climate, consistent year-round sea temperatures above 20 °C, and waters laden with natural feed. In addition to the suitable environment, the native *P. vannamei* of this region is highly suited to pond rearing (Wyban and Sweeney 1991). Over time, these shrimp have become the primary pond-reared shrimp species in the Americas and expanded globally; more than half of all crustaceans reared in pond environments are now *P. vannamei* (UN FAO Fisheries and Aquaculture Department 2018c).

P. vannamei has numerous properties that make it desirable for farming compared to other shrimp species (Table 2.1). It grows rapidly; is amenable to high stocking densities as it is not as aggressive as other species; has a relatively low protein requirements; is tolerant of lower water quality and a wide range of salini-

Table 2.1 Characteristics of *P. vannamei* that make them suitable for pond culture (Wyban and Sweeney 1991)

P. vannamei	Characteristic
Growth rate	1–1.5 g per week up to 20 g.
Stocking density	Typical stocking densities 60 m^3–150 m^3, up to 400 m^3.
Aggressiveness in ponds	Not as aggressive as other farmed species.
Salinity tolerance	They tolerate a wide range of salinities from 0.5 ppt to45 ppt.
Larval rearing	High survival rates in hatchery of 50–60%.
Post-harvest	If treated with ice, they will not discolor.

ties; resists melanosis[1]; and has a taste and color that is agreeable to United States shrimp consumers (Briggs et al. 2004). Also, captive *P. vannamei* females have been known to produce as many as 2.2 million eggs[2] during their lifespan (Wyban and Sweeney 1991). For these reasons and the fact that it is native to Ecuador, *P. vannamei* is only shrimp species reared in all Ecuadorian shrimp farms.

Examining the *P. vannamei* lifecycle is vital to understanding both their preferred use in aquaculture and their reliance on the mangrove forest. In the postlarval stage, *P. vannamei* molt every 4–6 days (Wyban and Sweeney 1991). During this vulnerable stage, *P. vannamei* generally burrows in the soft detritus of the mangrove forest environment to evade predators. The mangrove forest also provides juvenile habitat. *P. vannamei* are carnivorous during the postlarval, estuarine/mangrove forest period of their life cycle, feeding on microorganisms in the estuary as well as detritus. In the natural environment, estuarine mangrove forests provide the feeding grounds and habitat that enable the transformation from postlarvae into full-grown adult shrimp before they return to the ocean for spawning.

2.5 The Environmental and Human Impacts of Ecuadorian Shrimp Farms

The environmental impacts of shrimp farming can be divided into two categories, the environmental damage and ecological disruption caused by the removal of mangrove forests and the environmental damage and ecological disruptions caused by day-to-day operations of the shrimp farms (Flaherty and Karnjanakesorn 1995). The single biggest negative environmental impact of shrimp farming in Ecuador is likely the loss of mangrove forests and the ecological functions they provide, as documented in this manuscript, but numerous other shrimp farming practices are a cause of environmental concern.

[1] Melanosis (Black Spot) is a harmless discoloration of shrimp. Consumers avoid shrimp with melanosis.

[2] The first larval stage of a crustacean lifecycle.

As early as 1989, Twilley (1989) documented the impacts of shrimp farms on the ecology of Guayas Estuary. The primary concern was the pumping and draining of estuarine waters into and out of shrimp ponds. Twilley (1989) state that pumping of estuarine water into the shrimp pond of Guayas Estuary in 1989 was 4×10^7 m^3 of water assuming 30,000 semi-intensive growout ponds and the amount of water exchanged is not only higher than the inputs of the Guayas River to the estuary during dry period, but it is also one-third of the freshwater input during the wet season. Such substantial movements of water have the potential to alter the entire geochemistry of the estuary. For example, evaporation rates of standing water in shrimp farms is higher than in the wider estuary, resulting in hypersaline waters flowing out of the ponds and entering the lower salinity estuary. This increase in the salinity of the estuary has implications for all fauna and flora that inhabit it. The water exiting farms is generally at a higher temperature than the estuary, and this again likely impacts flora and fauna of the estuary. Also, changing the flow, salinity, and temperature of the estuary, growout ponds likely seep more saltwater into the terrestrial ground than the wider estuary does, as they are often located on the edge of the estuary or just outside the estuary, and this potentially increases the salinity of the groundwater. Indeed, the loss of farmland due to the increase in groundwater salinity is an environmental problem noted in Ecuador that has resulted from the expansion of shrimp farms out of the estuary into the nearby terrestrial environment (Snedaker 1986).

Shrimp ponds drain into the surrounding estuary on an almost constant basis. Besides salinity and temperature, the contents of the water draining into the estuary are a point of environmental concern. The major issues are related to shrimp farm inputs, mostly from the external feed, and shrimp farm nutrient outflow, primarily from shrimp waste. A high proportion of the nutrients in shrimp feed end up as waste, and this waste is discharged into the estuary in concentrated levels which can cause a rapid deterioration in water quality near the outflow of the shrimp ponds. During the transition from one shrimp crop to another, the wastewater discharge can contain as much as one ton of organic waste per hectare of pond (Flaherty and Karnjanakesorn 1995). The waste contains nutrients, organic solids, chemicals, and bacteria (Lin 1993).

Shrimp ponds contain higher levels of nutrients, biological oxygen demand (BOD), and salinity than estuarine waters outside the ponds. The effluent load discharged annually from farms in Chone Estuary alone is estimated to equal the domestic waste load of 1.5–2.5 million human inhabitants (Arriaga et al. 1999). The critical threshold, as it pertains to estuarine health, appears to be between 40% and 60% conversion of mangrove forest to shrimp farm within the estuary. Stram et al. (2005), notes that when 40% of the mangrove forest in the upper estuary or 60% in the lower estuary are lost to shrimp farming, dissolved inorganic nitrogen levels approach the same level as they would if 100% of the mangrove forest had been converted to shrimp ponds. Such high levels of nutrients emanating from shrimp ponds have resulted in numerous harmful algal blooms (HABs) along the coast of Ecuador that are deadly to numerous fish and invertebrate species, including wild shrimp and farmed shrimp in the ponds (Twilley 1989; Jimenez 1989; Stram et al. 2005).

Additional ecological concerns about shrimp farming relate to the escape of exotic and non-native shrimp species that are often used in aquaculture (Tobey et al. 1998). Though this is not a concern in Ecuador as the farmed-shrimp are a native species, non-native species pond escape has occurred in other aquaculture systems globally, including *P. vannamei* on the east coast of the United States. Another concern about shrimp farming is that diseases related to shrimp farming and high stocking densities, such as TSV, WSSV, vibriosis, or new diseases that may occur can develop in the shrimp ponds and infect the wild shrimp population, but this has not yet occurred in Ecuador (Tobey et al. 1998). An early and significant ecological concern of the shrimp farm industry in Ecuador was the potential impact on wild shrimp populations of extracting postlarvae and egg-laden female shrimp from the estuaries (Clay 1996; Snedaker 1986). This concern has now been substantially mitigated in Ecuador by the development and use of broodstock in the shrimp hatcheries. The presence of growout pond dikes and infrastructure in densely farmed environments, as in Fig. 2.3, and other infrastructure likely alters the hydrologic conditions of the estuary potentially isolating mangrove forest communities from brackish water, altering the tidal inundation regime of the estuary, and potentially inundating mangrove forests with highly saline water when it is discharged from the shrimp ponds (Tobey et al. 1998). Lastly, the shrimp farm dikes are prone to erosion and likely lead to increased levels of sedimentation in estuaries that are already often suffering from issues of poor water quality due to increases in sedimentation from terrestrial sources (Leon et al. 2016; Hamilton et al. 2015; Stram et al. 2005; Coello et al. 1993).

A major shrimp farm environmental issue raised by traditional fishers in Chone Estuary during the author's time working in that region and raised by nearby residents to Collins (2010) is the use of poison to kill fish in the ponds before the postlarvae arrive. This poison is then released into the estuary, killing wild fish. After much investigation, the author did confirm that the early shrimp farm best practices recommend the use of rotenone to kill fish before the postlarvae arrive (Cook and Rabanal 1978). Rotenone is a broad-spectrum insecticide, piscicide, and pesticide particularly useful for killing fish but leaving other species alive. In South America, rotenone can be extracted from the roots of numerous tree and plant species (Krukoff and Smith 1937), and it is known to be harvested on Isla Puná for use as a piscicide in the shrimp farm industry (Krukoff and Smith 1937).

It is not only in the area of environmental health that aquaculture is a concern, but human health concerns are now being realized. Sapkota et al. (2008) elucidate eight pathways from aquaculture to human health concerns including issues around antibiotics, wastewater, metals, organohalogens, other agrochemicals, and transgenes. Antibiotic use is a long-term and well-known issue in shrimp aquaculture in Ecuador as a means of preventing vibriosis. Sapkota et al. (2008) point to 13 peer-reviewed studies that show elevated levels of antibiotic resistance in and around aquaculture ponds including studies around shrimp ponds. They conclude that the data presented across a range of studies support the theory that the development of bacterial antibiotic resistance in aquaculture environments could contribute to or influence bacterial antibiotic resistance in human populations.

After antibiotic use and resistance, the second human health concern with farmed shrimp is ingestion of halogenated organic compounds resulting from the consumption of aquaculture food products. These compounds include the cancer-causing dioxins, polychlorinated dibenzodioxins (PCDDs) and polychlorinated dibenzofurans (PCDFs), as well as polychlorinated biphenyls (PCBs), and polybrominated diphenyl ethers (PBDEs) (Sapkota et al. 2008). A United States-based food advocacy group listed nine halogenated organic compounds present in shrimp aquaculture feed that are either banned or labeled as not fit for human consumption (Public Citizen's Food Program 2004). It is not just in the advocacy arena that halogenated organic compounds in aquaculture feed supply are a concern. In a literature review on the subject, Sapkota et al. (2008) find at least three studies where the halogenated organic compounds present in the tissue of aquaculture species can be traced back to the input feed. Other human-health concerns relate to human excreta and wastewater use in aquaculture that remains common in Asia (World Health Organization 2006) but is not practiced in shrimp aquaculture in Ecuador, although accidental seepage into ponds from human waste systems may occur.

Shrimp aquaculture in Ecuador likely contributes to HABs across all the studied estuaries. HABs have been shown to have numerous human health consequences (Berdalet et al. 2016) including neurotoxic shellfish poisoning, inhalation exposure, and a new potential illness known as brevetoxin fish poisoning (Fleming et al. 2011). Indeed, five toxic syndromes have been identified among humans who have eaten HAB-exposed shellfish, and these conditions can result in severe illness or death (James et al. 2010). The study of the human health and ecological impacts of HABs in Ecuador is in early stages, but research into the connection and the human perception of the connection is ongoing (Borbor-Córdova et al. 2018).

In addition to the environmental and human health impacts of aquaculture, there are social implications of shrimp farming. Indeed, the socioeconomic issues resulting from shrimp farm expansion in Ecuador likely outweigh the technical and management issues resulting from shrimp farming (Snedaker 1986). Many of these issues are covered in Chap. 3.

2.6 Ecuadorian Shrimp Farm Expansion and the Legal Framework of Ecuadorian Shrimp Farms

As the estuarine environments are defined by the extreme margins of the highest tides, all lands within them have historically been the responsibility of the Ecuadorian Navy and under state ownership. This likely means that even the first few shrimp farms to appear in the estuary have always been illegal. Since the Decree 2939-B in 1978, destruction of mangrove forests for aquaculture pond construction has been illegal (Robadue 1995). As 99% of shrimp farms appeared after 1978 (Parks and Bonifaz 1994; CLIRSEN 2007), any that were built in mangrove forests are technically illegal. The 1978 decree was reinforced in 1985 with Decree 825, and the accompanying fisheries law which not only prohibited the removal or alteration of mangrove forests for shrimp farm construction but additionally obligated shrimp farm owners to protect the land where mangrove forests remained (Robadue 1995).

As with the decrees covering mangrove forest conservation covered in Chap. 3, a lack of enforcement, knowledge of the law, and the drive for aquaculture development locally, nationally, and globally resulted in the shrimp farm laws having no meaning on the ground, and shrimp farm expansion, particularly at mangrove forests' expense, continued throughout this period. As a result, in 1979 the entirety of Ecuador had 5416 ha of shrimp farms, by 1984 this had increased to 89,400 ha, by 1987 it was 117,700 ha, and by 1991 it was 146,00 ha (Parks and Bonifaz 1994; CLIRSEN 2007). By 1999 the shrimp farm area estimate for Ecuador was 175,167 (CLIRSEN 2007).

2.7 Aquaculture and Shrimp Farm Summary

It is incontestable that aquaculture and shrimp farming has grown at remarkable rates in developing countries over the last 40 years. Use of water resources to raise seafood, as opposed to merely catching seafood, mirrors processes that have been and are occurring in the terrestrial food supply systems for tens of thousands of years. As opposed to land-based farming practices, or even offshore deep-water aquaculture, preferred shrimp farming environments are available in a highly limited geographical setting in tropical estuaries with the correct environmental conditions.

Shrimp farms and the broader aquaculture industry now provide significant returns to local and national economies in Ecuador and elsewhere in the tropical developing world. Despite the economic returns, protein supply, and livelihood options offered by shrimp farming, questions remain about the long-term economic viability, livelihood benefits, and food security offered by the shrimp farm industry. Most of the questions raised about the long-term benefits of shrimp farms are based on the fact that shrimp farms may displace highly productive mangrove forest ecosystems that are themselves drivers of livelihoods, food security, and high-quality protein production in the local communities. Additional questions are raised as to the role shrimp farming plays in changing the estuarine environment on which residents rely, and the fact that shrimp aquaculture itself tends to be a boom and bust industry with ponds abandoned after short periods due to disease and the numerous environmental and human health implications of shrimp farming.

International aid, international financing institutions, and multi-national corporations directly financed the nascent shrimp farming industry in Ecuador. Such institutions are generally viewed as providing aid, development assistance, or investments that benefit the receiving nation or region. Doubts about the potential benefits of aquaculture call into question the motives and practices behind such financing. More specifically, who benefited from shrimp farm financing? Did the local communities, the wider regional or national economy, or the developed economies of the world?

The methods employed by shrimp aquaculture during the shrimp lifecycle, the shrimp species utilized in the farming system, the feed and other inputs utilized by shrimp farms, and most importantly, the location of growout ponds may all have broader ecological and human health consequences that are not fully accounted for within the existing economic and environmental analyses of commercial shrimp farming. Nature provides numerous subsidies to shrimp farming, including clean water to maintain the growout ponds, processing of waste and other chemicals exported from the farms, fishmeal-derived feed for the shrimp, and perhaps most importantly, a nutrient-rich estuarine environment for the growout ponds. Without accounting for such natural subsidies, the actual true and complete cost/benefit analysis of shrimp farming remains incomplete.

As already noted, the single biggest concern surrounding shrimp farming is the amount of mangrove forest it displaces. The sheltered estuaries of Ecuador that shrimp farms now dominate were the historic homes of most of the mangrove forests in Ecuador. This manuscript not only attempts to quantify mangrove forest loss and shrimp farm expansion in these environments since the arrival of shrimp farms but additionally attempts to answer the question of how many of the shrimp farms directly displaced mangrove forest? This displacement question is at the heart of the debate over the merits of shrimp aquaculture in Ecuador (Twilley 1989).

The scale of displacement of large amounts of mangrove forest by shrimp ponds or other forms of aquaculture is disputed. For example, Menasveta (1997) states that the majority of aquaculture systems are located above the high tide level and hence, away from mangrove forests. Famously in the environmental literature, Boyd and Clay (1998) state that less than 10% of the total global mangrove forests have been lost to aquaculture. This statement has propagated throughout the environmental literature and beyond, and in many cases, it is misused to diminish the negative environmental impacts of shrimp ponds, but this 10% estimate has been demonstrated to be a misunderstood figure that is both spatially and temporally problematic (Hamilton 2015). The direct aquaculture to shrimp conversion number is likely far higher, both in Ecuador and globally (Hamilton 2013).

Recommended Further Reading

The P. *vannamei* Lifecycle and Aquaculture Process.

UN FAO (2014)

The Environmental Impacts of Shrimp Farming in Ecuador

Twilley (1989)
Tobey et al. (1998)

Human Health Concerns of Shrimp Farming and HABs

Sapkota et al. (2008)
Fleming et al. (2011)

References

Arriaga L, Montaño M, Vásconez J (1999) Integrated management perspectives of the Bahia de Caráquez zone and Chone River estuary, Ecuador. Ocean Coast Manag 42(2–4):229–241. https://doi.org/10.1016/s0964-5691(98)00055-6

Asian Development Bank (1991) Global industry update: shrimp. Global industry updates, Kuala Lumpur

Berdalet E, Fleming LE, Gowen R, Davidson K, Hess P, Backer LC, Moore SK, Hoagland P, Enevoldsen H (2016) Marine harmful algal blooms, human health and wellbeing: challenges and opportunities in the 21st century. J Mar Biol Assoc U K 96(1):61–91

Borbor-Córdova MJ, Pozo-Cajas M, Cedeno-Montesdeoca A, Mantilla Saltos G, Kislik C, Espinoza-Celi ME, Lira R, Ruiz-Barzola O, Torres G (2018) Risk perception of coastal communities and authorities on harmful algal blooms in Ecuador. Front Mar Sci 5(365). https://doi.org/10.3389/fmars.2018.00365

Boyd CE, Clay JW (1998) Shrimp aquaculture and the environment. Sci Am June:58–65

Briggs M, Funge-Smith S, Subasinghe R, Philips M (2004) Introductions and Movement of *Penaeus Vannamei* and *Penaeus Stylirostris* in Asia and the Pacific. FAO, Regional Office for Asia and the Pacific, Bangkon

Cheshire J (2005) Memoirs of a shrimp farmer: The story of how a giant new industry developed. Voyd Brothers, INC., Panama City

Clay JW (1996) Market potentials for redressing the environmental impact of wild captured and pond produced shrimp. sn, Sl

CLIRSEN (2007) Actualizacion Del Estudio Multitemporal de Manglares, Camaroneras Y Areas Salinas En Las Costa Continental Ecuatoriana Al Ano 2006. vol 1. Centro De Levantamientos Integrados De Recursos Naturales Por Sensores Remotos, Quito, Ecuador

Coello S, Proafio-Lerowr D, Robadue D (1993) Special area management planning in Ecuador's Rio Chone estuary. Paper presented at the coastal zone, New Orleans, LA, July

Collins S (2010) Mangrove destruction and shrimp aquaculture in Ecuador: a focus on property right enforcement. University of Ottawa, Ottawa

Cook HL, Rabanal H (1978) Manual on pond culture of penaeid shrimp. ASEAN National Coordinating Agency of the Philippines and FAO/UNDP, Manila. South China Sea Fisheries Development and Coordinating Programme, Manila. ASEAN National Coordinating Agency of the Philippines, Manila, Philippines

Diana JS (2009) Aquaculture production and biodiversity conservation. Bioscience 59(1):27–38. https://doi.org/10.1525/bio.2009.59.1.7

Flaherty M, Karnjanakesorn C (1995) Marine shrimp aquaculture and natural resource degradation in Thailand. Environ Manag 19(1):27–37. https://doi.org/10.1007/BF02472001

Fleming LE, Kirkpatrick B, Backer LC, Walsh CJ, Nierenberg K, Clark J, Reich A, Hollenbeck J, Benson J, Cheng YS (2011) Review of Florida red tide and human health effects. Harmful Algae 10(2):224–233

Gautier D, Amador J, Newmark F (2001) The use of mangrove wetland as a biofilter to treat shrimp pond effluents: preliminary results of an experiment on the Caribbean coast of Columbia. Aquac Res 32:787–799. https://doi.org/10.2307/2389096

Hamilton SE (2012) The impact of shrimp farming on mangrove ecosystems and local livelihoods along the Pacific Coast of Ecuador, vol 1. Dept. of Geography and Geology, Doctoral Dissertation, vol 3477166, 1st edn. ProQuest, UMI Dissertation Publishing, Hattiesburg, Mississippi

Hamilton SE (2013) Assessing the role of commercial aquaculture in displacing mangrove forest. Bull Mar Sci 89(2):585–601. https://doi.org/10.5343/bms.2012.1069

Hamilton SE (2015) No evidence that shrimp aquaculture is responsible for minimal mangrove deforestation. Bioscience 65(5):457–457. https://doi.org/10.1093/biosci/biv030

Hamilton SE, Stankwitz C (2012) Examining the relationship between international aid and mangrove deforestation in coastal Ecuador from 1970 to 2006. J Land Use Sci 7(2):177–202. https://doi.org/10.1080/1747423x.2010.550694

Hamilton SE, Fitt J, Kaempf C, Coyne SJ (2015) The impacts of sedimentation in the Río Chone estuary. In: Toulkeridis T, Andrade L, Mejía M, Cartay R, Regalado L (eds) VIII Jornadas Académicas Turismo y Patrimonio, Compartiendo lo nuestro con el mundo. Memorias Contribuciones Científicas., vol 1. Descubriendo el Geoturismo en Manabí: Distrito Jama – Pedernales, vol 1. ESPAM MFL, Calceta, pp 38–43

Hogarth PJ (2015) The biology of mangroves and seagrasses. Oxford University Press, Oxford

Hopkins JS, Sandifer PA, DeVoe MR, Holland AF, Browdy CL, Stokes AD (1995) Environmental impacts of shrimp farming with special reference to the situation in the continental United States. Estuaries 18(1):25–42

Idyll CP (1965) Shrimp nursery: science explores new ways to farm the sea, vol 127. National Geographic

James K, Carey B, O'halloran J, Škrabáková Z (2010) Shellfish toxicity: human health implications of marine algal toxins. Epidemiol Infect 138(7):927–940

Jimenez R (1989) Red tide and shrimp activity in Ecuador. Establishing a sustainable shrimp mariculture industry in Ecuador, Technical Report Series TR-E-6. Coastal Resources Center, University of Rhode Island, Narragansett

Krukoff BA, Smith AC (1937) Rotenone-yielding plants of South America. Am J Bot 24(9):573–587. https://doi.org/10.2307/2436635

Leon FD, Johnson S, Marcon J, Phares N (2016) Watershed management to address sedimentation of Ecuador's Rio Chone ESTUARY. University of California, Santa Barbara

Lin C (1993) Resource recovery from wastewater of intensive shrimp farming. In: Proceeding of fisheries seminar on "positive impact of shrimp industry to environmental, coastal quality and socio-economic in Thailand, Bangkok, Thailand, 1993, vol 1993. Department of Fisheries Ministry of Agriculture and Cooperatives, pp 126–136

Louis Berger Group T (2006) Business models for aquaculture in Iraq: IZDIHAR – USAID contract #267-C-00-04-00435-00. USAID, Washington, DC

McClennen C (2004) White spot syndrome virus: the economic environmental and technical implications on the development of Latin American shrimp farming. Law and Diplomacy, Tufts University, Medford

Menasveta P (1997) Mangrove destruction and shrimp culture systems. World Aquacult 28:36–42

MSINGI (2010) MSINGI: aquaculture. MSINGI. http://www.msingi.com/index.php/our-industries. Accessed 2 Jan 2019

Nash C (1987) ADCP/REP/87/25 – future economic outlook for aquaculture and related assistance needs. FAO, Rome

Naylor RL, Goldburg RJ, Mooney H, Beveridge M, Clay J, Folke C, Kautsky N, Lubchenco J, Primavera J, Williams M (1998) Nature's subsidies to shrimp and salmon farming. Science 282(5390):883–884

Paez-Osuna F (2001) The environmental impact of shrimp aquaculture: causes, effects, and mitigating alternatives. Environ Manag 28(1):10

Parks PJ, Bonifaz M (1994) Nonsustainable use of renewable resources: mangrove deforestation and Mariculture in Ecuador. Mar Resour Econ 9(1):1–18

Public Citizen's Food Program (2004) Chemical cocktail: the health impacts of eating farm-raised shrimp. Public Citizen, Washington, DC

Public Citizen's Food Program (2005) Fishy currency: how international finance institutions fund shrimp farms. Public Citizen, Washington, DC

Rajitha K, Mukherjee CK, Chandran VR (2007) Applications of remote sensing and GIS for sustainable Management of Shrimp Culture in India. Aquacult Eng 36(1):1–17

Ricque-Marie D, Abdo-de La Parra MI, Cruz-Suárez LE, Cuzon G, Cousin M, Pike IH (1998) Raw material freshness, a quality criterion for fish meal fed to shrimp. Aquaculture 165(1–2):95–109

Robadue DD (1995) Eight years in Ecuador: the road to integrated coastal management, vol 2088. Coastal Resources Center, University of Rhode Island, Narragansett

Rodríguez FVL (2018) Mangrove concessions: an innovative strategy for community mangrove conservation in Ecuador. In: Makowski C, Finkl CW (eds) Threats to mangrove forests: hazards, vulnerability, and management, vol 25. Springer, Cham, pp 557–578

Rodríguez J, Bayot B, Amano Y, Panchana F, de Blas I, Alday V, Calderón J (2003) White spot syndrome virus infection in cultured *Penaeus Vnnamei* (Boone) in Ecuador with emphasis on histopathology and ultrastructure. J Fish Dis 26(8):439–450

Rosenberry B (2008) Dennis Zensen: inspired the development of shrimp farming in the Western Hemisphere. Shrimp News International, vol January 18, 2008. Shrimp News International, San Diego

Sapkota A, Sapkota AR, Kucharski M, Burke J, McKenzie S, Walker P, Lawrence R (2008) Aquaculture practices and potential human health risks: current knowledge and future priorities. Environ Int 34(8):1215–1226

Shang YC (1992) Penaeid markets and economics. In: Marine shrimp culture. Elsevier, Amsterdam, pp 589–604

Shehadeh Z, Orzeszk J (1997) Selected developments and trends: external assistance. In: Orzeszk J, Maclean J (eds) Review of the state of world aquaculture. FAO, Rome

Skladany M, Harris C (1995) On global pond: international development and commodity chains in the shrimp industry. In: McMichael P (ed) Food and agrarian orders in the world-economy. Praeger, Westport, pp 169–195

Snedaker SC (1986) Traditional uses of South American mangrove resources and the socio-economic effect of ecosystem changes. In: Kunstadter P, Bird ECF, Sabhasri S (eds) Workshop on man in the mangroves, vol 1. United Nations University, Tokyo, pp 102–112

Stram DL, Kincaid CR, Campbell DE (2005) Water quality modeling in the Rio Chone estuary. J Coast Res 21(4):797–810. https://doi.org/10.2112/011-NIS.1

Tacon AG, Metian M (2008) Global overview on the use of fish meal and fish oil in industrially compounded aquafeeds: trends and future prospects. Aquaculture 285(1–4):146–158

Tobey J, Clay J, Vergne P (1998) Maintaining a balance: the economic, environmental and social impacts of shrimp farming in Latin America, Coastal resource management project II. University of Rhode Island, Kingston

Tomlinson PB (1986) The botany of mangroves. Cambridge University Press, Cambridge

Transboundary Shrimp Viral Diseases with Emphasis on White Spot Syndrome Virus (WSSV) and Taura Syndrome Virus (TSV) (2004) The Government of Japan Trust Fund: Regional Fish Disease Project. http://rfdp.seafdec.org.ph/publication/proceedings.html. Accessed 23 June 2004

Twilley RR (1989) Impacts of shrimp mariculture practices on the ecology of coastal systems in Ecuador. Establishing a sustainable shrimp mariculture industry in Ecuador, vol Technical Report Series TR-E-6. Coastal Resources Center, University of Rhode Island, Narragansett

UN FAO (2014) Penaeus vannamei (Boone, 1931) (trans: Department FaA). Cultured aquatic species information Programme. UN FAO Fisheries and Aquaculture Department, Rome, Italy

UN FAO Fisheries and Aquaculture Department (2006) State of world fisheries and aquaculture. State of world fisheries and aquaculture. Food & Agriculture Organization of the UN, Rome. doi:978-9251051771

UN FAO Fisheries and Aquaculture Department (2017) FishStat plus – universal software for fishery statistical time series. FAO, Rome

UN FAO Fisheries and Aquaculture Department (2018a) Feed formulation. Aquaculture feed and fertilizer resources information system. UN FAO Fisheries and Aquaculture Department, Rome

UN FAO Fisheries and Aquaculture Department (2018b) Species fact sheets: *Penaeus vannamei*. The state of world fisheries and aquaculture. UN FAO Fisheries and Aquaculture Department, Rome

UN FAO Fisheries and Aquaculture Department (2018c) The state of world fisheries and aquaculture 2018—meeting the sustainable development goals. The state of world fisheries and aquaculture. UN FAO Fisheries and Aquaculture Department, Rome

United Nations Committee on World Food Security (2003) The role of aquaculture in improving food security and nutrition. Paper presented at the committee on world food security 29th session, Rome, Italy, May 12–16, 2003

USAID (2003) Ecuador northern border development: 518–013. United States Agency for International Development, Washington, DC

Venizelos A, Benetti DD (1996) Epitheliocystis disease in cultured yellowtail *Seriola mazatlana* in Ecuador. J World Aquacult Soc 27(2):223–227

Wood A, Mayer J (2007) Changing the face of the waters: the promise and challenge of sustainable aquaculture. World Bank, Washington, DC

World Bank (2017) World integrated trade solution: trade summary for Ecuador. World Bank, New York

World Health Organization (2006) Guidelines for the safe use of wastewater, excreta and greywater, vol 1, 2, 3, 4. World Health Organization, Geneva

Wyban J, Sweeney JN (1991) Intensive shrimp production technology: the oceanic institute shrimp manual. The Oceanic Institute, Honolulu

Chapter 3
Introduction to Coastal Ecuador

Abstract This chapter delineates and describes each of the study sites in Ecuador. The chapter operates from north to south beginning with the estuary of Cayapas-Mataje estuary in the north and ending with the estuaries of El Oro province in the south. Detailed maps are presented for each study area that delineates the boundaries of the analysis and represents all the protected areas within each estuary. For each of the eight delineated estuaries, the following are described; the protected status, the climate, the environmental conditions, the prevailing socioeconomic conditions, the local livelihoods, and the estuarine biodiversity. Finally, the legal framework of both shrimp farms and mangrove forest are examined. In addition to the facts and figures of the study areas, this chapter attempts to represent the milieu of the local residents of each estuary with a focus on both Chone Estuary and Cayapas-Mataje Estuary.

Keywords Ecuador · Coastal Ecuador · Estuaries · Cayapas-Mataje · Muisne · Cojimies · Chone · Isla Puná · Guayas · El Oro

3.1 The Estuaries of Coastal Ecuador

As depicted in Fig. 3.1, only the four Ecuadorian provinces of Esmeraldas, Manabí, Guayas, and El Oro border the Pacific Ocean, and together they form the Costa region of Ecuador. Due to its lowland location, Los Ríos Province is classified as part of the Costa, but it has little or no mangrove forest because it is far from any saltwater environment. The province of Galápagos does have a Pacific coastline and limited mangrove forest cover, but it is not included in this analysis which focuses only on mainland Ecuador. The research area for this study comprises all the major estuaries within these four coastal provinces capable of sustaining mangrove forests

---

Electronic Supplementary Material The online version of this chapter (https://doi.org/10.1007/978-3-030-22240-6_3) contains supplementary material, which is available to authorized users.

S. E. Hamilton, *Mangroves and Aquaculture*, Coastal Research Library 33,
https://doi.org/10.1007/978-3-030-22240-6_3

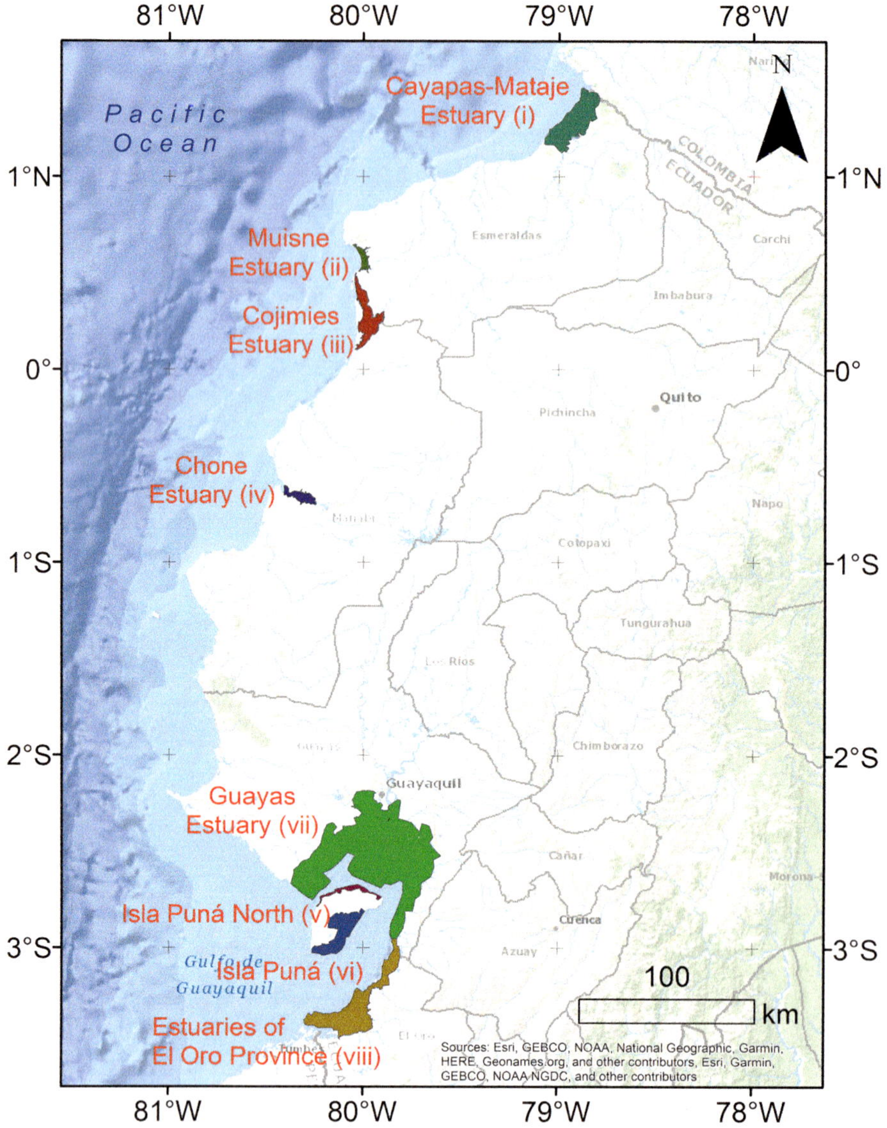

Fig. 3.1 Ecuador study sites

and suitable for commercial shrimp farming. As opposed to other analyses, we do not divide contiguous estuaries at provincial borders; we attempt to isolate each estuary and conduct the analysis at this estuarine level. By aggregating the estuaries to provincial boundaries as necessary, we find the estuary to be the observable unit, and neither the mangrove forests nor residents pay much heed to the boundaries that pass through them. The estuarine environments included from north to south are (Fig. 3.1):

(i) Cayapas-Mataje Estuary (REMACAN)
(ii) Muisne Estuary
(iii) Cojimíes Estuary
(iv) Chone Estuary
(v) Isla Puná North
(vi) Isla Puná
(vii) Guayas Estuary
(viii) The estuaries of El Oro Province

In Esmeraldas province, two major estuaries are included. The most northern and pristine is the (i) Cayapas-Mataje Estuary in and around the Cayapas-Mataje Mangrove Reserve (REMACAN) located on the Colombian border. The second is (ii) Muisne Estuary, located around the city of Muisne, farther south in Esmeraldas Province. In Manabí Province, the study areas are (iii) Cojimíes Estuary on the northern provincial border with Esmeraldas, and (iv) the Chone Estuary in and around the city of Bahía de Caráquez. The provincial border runs directly through the center of Cojimíes Estuary. The study areas for Guayas province are (v) Isla Puná North, (vi) Isla Puná, and (vii) the Guayas mainland in and around the Guayas River. In El Oro Province, the study area is the entire coastline of (vii) El Oro province including Grande Estuary in the south and numerous creeks and rivers to the north. These regions account for approximately 99% of the 1969 base level mangrove forest present in Ecuador (Chap. 1, Table 1.2) (CLIRSEN 1987, 2007).

To assist in delineating the study areas at high granularity, it was assumed that nowhere with an elevation 6 m receives tidal saltwater input, and therefore, these areas will not support coastal mangrove forests. The 6 m delineation is based on a maximum tidal range in our study areas of 4 m (Blanchard and Prado 1995), amplified to a maximum of 5.29 m (Bodero 1993). The extension to 6 m allows for localized regional topography that may magnify the estuarine tidal ranges beyond the maximum reported value. These areas must be accounted for because mangrove forests may exist in these interior and occasionally flooded parts of the estuary. To obtain all of the topography below 6 m, a 30 m resolution digital elevation model was derived using elevation data from the Instituto Geográfico Militar (IGM) topographic database of 30 million elevation points within Ecuador (Souris 2008). By extracting all areas with elevations below 6 m, we were able to narrow the potential mangrove forest area down to a more limited area. Final delineation of the study sites was conducted by utilizing topographic maps to identify the major estuaries within these parameters. The study area delineation process returned eight estuaries of which seven are distinct. The Guayas study area (vii) and the El Oro study areas (viii) are likely one single estuarine environment but are divided into two for this analysis.

3.2 Cayapas-Mataje Estuary and the Cayapas-Mataje Ecological Mangrove Reserve (REMACAN): An Artisanal Fishing Community in Action

Figure 3.2 depicts the study area (Fig. 3.1, i), referred to in this manuscript as Cayapas-Mataje or REMACAN. Cayapas-Mataje is located on the Colombian border in northern Esmeraldas Province in and around REMACAN at the confluence of the Cayapas, Mataje, Santiago, and San Miguel Rivers. The study area is50,625 ha of which 48,556 is preserved. The region is considered to be Ecuador's most pristine mangrove forest environment (Ocampo-Thomason 2006) and potentially the most pristine along the entire Pacific coast of the Americas (Wetlands International 2004). Mangrove forest is the dominant landcover, and Cayapas-Mataje contains the tallest known mangrove trees in the world, with heights up to 64 m (Spalding et al. 2010) and diameters over 1 m (Hamilton et al. 2017). The dominant mangrove is *Rhizophora mangle*, which constitutes 98% of all the mangrove forest area. The average *Rhizophora mangle* individual tree in Cayapas-Mataje has a diameter at breast height (DBH) that is approximately double that of the other estuaries nearby (Hamilton et al. 2017), and amongst them is potentially the world's tallest mangrove tree, measuring greater than 60 m in height (Ochoa 1997). Indeed, the average *Rhizophora mangle* height here is greater than 40 m (Hamilton et al. 2017). These mangrove forests exist in a region consisting of pristine estuarine environments,

Fig. 3.2 Cayapas-Mataje estuary and REMACAN

pristine freshwater and inter-tidal flooded wooded wetlands, and pristine wooded peatlands (Wetlands International 2004).

3.2.1 Climatic & Environmental Conditions

The water temperature in and around Cayapas-Mataje is a consistent 26–29 °C year-round. This is the smallest thermal fluctuation in all of Ecuador (Ministerio del Ambiente 2018) as this region is not impacted by the cooler Peruvian/Humboldt Current and is permanently in the grasp of the warm equatorial counter current (Chap. 1, Fig. 1.4). This region has the highest rainfall of all of coastal Ecuador, with annual rainfall totals averaging 3000–4000 mm (Bodero 1993) with a very limited and somewhat drier period in June and July. Aside from this short dry season, the area is generally overcast. The air temperature varies little from month to month across the year with high temperatures hitting 29 °C almost daily and low temperatures rarely dropping far below 23 °C. The Köppen-Geiger climate classification is Af, or Equatorial and full-humid (Kottek et al. 2006). The entire study site has no elevations higher than a couple of meters.

3.2.2 Socioeconomic Conditions & Livelihoods

Almost all of the Afro-Ecuadorian population living in and around the 44,000 km^2 reserve area and the surrounding towns rely on the mangrove forest for their income, with over 85% of households supplementing their income from the traditional uses of the forests (Ocampo-Thomason 2006). The populations of Cayapas-Mataje collect concha (Fig. 3.3, Panel 4), clams, shells, oysters, crabs (Fig. 3.3, Panel 3), shrimp, fish, and prawns (Ministerio del Ambiente 2018) and appear to live an artisanal lifestyle almost wholly supported by the mangrove forest. As of the late 1990s, the estuary sustained 4000 concha harvesters, called concheras (MacKenzie Jr. 2001). The primary form of transport around the numerous islands is dugout canoes equipped with a small sail (Fig. 3.3, Panel 2). Homes are most often built on stilts, and the roads are inundated daily at high tide (Fig. 3.3, Panel 1). People live in mangrove forest wood homes across the numerous islands within Cayapas-Mataje (Fig. 3.3, Panel 1). Outside of the main road into the town, no roads or bridges connect the isolated communities. The major hub of activity is the town dock (Fig. 3.3, Panels 3 and 4) from which people move to and from the islands, and the fishers and concheras leave in the morning and return with their catch at night. The 2008 population estimate within 5 km of the entire estuary is 17,104, and the population estimate within 10 km of the entire estuary is 21,677 (Oak Ridge National Laboratory and UT-Battelle LLC 2014; Hamilton 2012), including residents on the Colombian side of the estuary, across which the local fishers appear to move freely.

Fig. 3.3 Cayapas-Mataje lifestyle & livelihoods
Caption – Panel 1 is the principal town of the region, San Lorenzo. The edge of town just inland from the town dock is pictured. The mangrove-constructed homes are often on stilts, and extensive tall mangroves are visible in the background. Panel 2 is a transport boat from San Lorenzo to one of the numerous islands with fishing boats in the background. Again, mangrove forest is visible in the background. The photograph is taken from another boat. Panel 3 is the crab catch at the end of the day. Each bag is a bushel of blue crab, and the picture is taken from where the town dock meets the main road. Panel 4 is the concha negra harvest being loaded onto the town dock at the end of the day to be ready for transport. Again, mangrove forests are present in the background. (All photographs are taken by the author)

The mangrove forests play an almost religious role in this region. Homes and businesses are adorned with mangrove forest murals (Fig. 3.4, Panel 1, 2, 3, and 4); Mangrove Day is celebrated across the community emphasizing the struggle to preserve the environment (Fig. 3.4, Panel 6); the town statue is dedicated to the fishers and concheras that work in the mangrove forest (Fig. 3.4, Panel 5); and the people of the region have successfully preserved their mangrove forest ecosystem and expelled the once numerous shrimp farms.

The role of women in the artisanal fisheries sector in Cayapas-Mataje cannot be over-emphasized. Almost all of the concha negra harvesters are women (Veach 1996; Ocampo-Thomason 2006), and in the town statue, the female figure is carrying a basket of concha negra (Fig. 3.4, Panel 5). Additionally, the women harvesting the concha negra among the *Rhizophora mangle* roots can be seen in the rear portion of Fig. 3.4, Panel 1. The local fishers, led by the female concheras, have organized

Fig. 3.4 Cayapas-Mataje community mangrove forest depictions
Caption – Panel 1 depicts a likely husband and wife pair catching fish in a net, with the women pulling in the net. In the background, numerous women are harvesting concha from the roots of the uniquely large *Rhizophora mangle*. Panel 2 depicts a festive event being held on the shore. Again, mangrove forests are visible in the background. Panel 3 is a typical bustling scene as the catch arrives. Notice the woman is bagging the catch. Again, mangrove forests are visible in the background. Panel 4 is the town dock from which the fisherfolk leave and the catch arrives. Again, mangrove forests are present in the background. Panel 5 is the town statute on the main road into town; an Afro-Ecuadorian fishing family is represented. The woman is carrying a basket of concha negra, the boy is carrying fish, and the man is carrying a clearing machete. Panel 6 is a community sign from an event celebrating the mangroves. The translation is, "Celebrate (Live) the Day of the Mangrove, community struggle and progress." (Photography by the author)

into a group named Fundación para la Defensa Ecológica (FUNDECOL) to resist shrimp farm expansion and protect the mangrove forest environment (Ocampo-Thomason 2006), and this group was pivotal in protecting the REMACAN site.

Aside from the mangrove forest economy, few other livelihood options exist in this region. San Lorenzo is a small commercial center with a few banks, restaurants, and hotels and a bustling town dock. The Navy and border police both maintain smalls compounds in the area and the National Park employs some local residents. Tourists used to visit the region by train from the Ecuadorian highlands from the late 1950s through the early 1990s, but since the tracks fell into disrepair, fewer tourists visit the region. Slowly but surely though, the train service is resuming in sections, and the Ecuadorian government is committed to returning the service to full functionality. Tourists also ceased visiting the area as it was and still is known to have a porous border with Colombia. During the Colombian conflict and occupation of the border region by the Marxist guerilla group, Revolutionary Armed Forces of Colombia (FARC), the region near Cayapas-Mataje was an area of high political instability with occasional kidnappings of international visitors. This region of coastal Ecuador has been historically isolated from Ecuador due to factors relating to political instability in the troublesome Colombian borderlands, thus leading to a lack of foreign investment.

A single road connects the Cayapas-Mataje to the Sierra via the town of Ibarra, but this road is subject to frequent closures due to rockslides. The primary coastal road, which connects Cayapas-Mataje to the provincial capital of Esmeraldas, was only connected to this region in recent years. This road now gives Cayapas-Mataje accessibility to the rest of coastal Ecuador including the important ports of Esmeraldas and Guayaquil. The United States Department of State historically had regular advisories against travel into this area due to drug activity and FARC violence.

3.2.3 Biodiversity

Numerous species inhabit the mangrove swampland. Cayapas-Mataje is home to 71 invertebrates and 329 vertebrates including 22 lizard species, 14 amphibian species, 68 fish species, 24 crustacean species, 17 gastropod species, and 29 bivalve species (Wetlands International 2004). The warm waters contain rays, sawfish, cod, sea-bass, mullet, snapper, and sargeant fish (Ministerio del Ambiente 2018) as well as abundant crab, lobster, langostino (squat lobster), shrimp, and the last viable concha negra population in Ecuador. Common birds spotted in the estuary include pelicans, frigates, cormorants, herons, and kingfishers. The mammal population includes the crab-eating raccoon, the nine-banded armadillo, ocelot, sloth, otter, agouti, anteater, white-tailed deer, and wild boar (Ministerio del Ambiente 2018). The American crocodile maintains a foothold in the park and is one of the last two remaining regions where it is commonly sighted in Ecuador.

3.2.4 *Protected Status*

As already noted, much of the area is protected as part of the REMACAN reserve which was set up to protect the pristine environment after the arrival of commercial aquaculture (Ministerio del Ambiente 2018). Cayapas-Mataje is unique as almost all of the study site is owned by the government and administered by the Ministry of the Environment with the express goal of supporting mangrove forest livelihoods and maintaining the pristine environment. The entire area is Ramsar site number 1292 incorporated in 2003 under the intergovernmental convention to protect wetlands (RAMSAR 2006).

3.3 Muisne Estuary and the Vida Silvestre Manglares of the Rio Muisne Estuary Refuge

Figure 3.5 depicts the study area (Fig. 3.1, ii), referred to in this manuscript as Muisne Estuary, which is in southern Esmeraldas Province. Muisne itself is a small tourist town on a riverine island, also known as Muisne. The study area is 5649 ha. The island is separated from the mainland by the Rio Muisne which technically flows both north and south of the island as well as on the eastern side (Fig. 3.2). The eastern, northern, and southern side of the island contain numerous small rivers that flow into the Rio Muisne, including the Rio Bunche at the extreme northern tip of the island; the Rio Vilsa and the Barro Estuary near the center of the island; and the Tortuga Estuary, Cuerval Estuary, and Rio Canuto in the south. The major remaining mangrove stands in this area were given protected status in 2003 as part of the Refugio de Vida Silvestre Manglares Estuario del Rio Muisne (Collins 2010), and this protection was enhanced in 2017 (Ministerio del Ambiente 2018). The remaining mangrove stands of this area are now subdivided into two areas, which themselves are further fragmented. The largest block of mangrove forest is in and around the communities of Ostiona, Isla El Congal, and Las Manchas; and the other set is further north in and around the communities of northern Isla de Muisne, Bunche, San Gregorio, San Francisco, and Bilsa (Ministerio del Ambiente 2018). The species composition is 71% *Rhizophora mangle*, 1% *Avicennia germinans*, and 28% *Laguncularia racemosa* (Hamilton et al. 2017).

3.3.1 *Climatic & Environmental Conditions*

Muisne Estuary likely has some limited influence from the cold Humboldt/Peruvian current (Chap. 1, Fig. 1.4), but the mangrove forests are in a well-sheltered estuary away from exposure to the ocean. The annual water temperature varies from a low of around 24 °C and a high of around 28 °C, with an annual average between 25 and

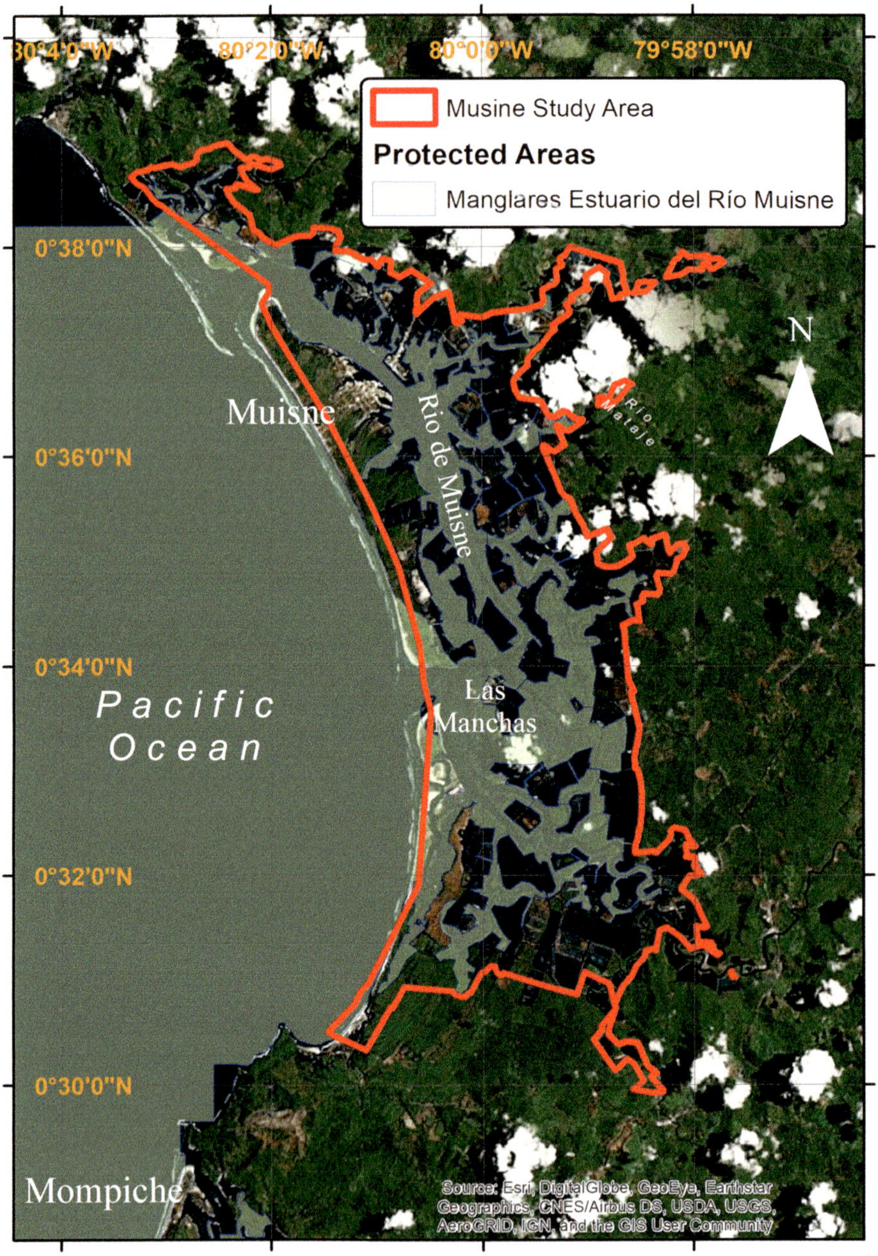

Fig. 3.5 Muisne estuary study area

26 °C. Muisne region occasionally has high levels of rainfall with annual rainfall totals averaging 500–3000 mm, with an average of about 1950 mm. Muisne has a pronounced dry season in June and July (Fick and Hijmans 2017). The highly variable rainfall totals are due to the southern oscillation., during El Niño years, the rainfall can be as much as ten times time higher than during a La Niña year, and as much as five times higher than the longer-term annual averages. The average monthly air temperatures show some seasonal variation but in general, the daily high temperatures are around 28 °C, and rarely drop below 22 °C (Fick and Hijmans 2017). The Köppen-Geiger climate classification is Aw, or Equatorial and winter-dry, but it is on the border of Af, or Equatorial and fully-humid (Kottek et al. 2006). The entire study site has no elevations higher than a couple of meters although some slight increase in elevation does exist to the north of the estuary but no more than 10 m.

3.3.2 Socioeconomic Conditions & Livelihoods

Muisne is majority Afro-Ecuadorian, but far less Afro-Ecuadorian than regions farther north. Indeed, Muisne may be the transition zone, with towns to the north having an Afro-Ecuadorian majority and towns to the south having Afro-Ecuadorians as the minority. Muisne has a mixed economy based on tourism, agriculture, and traditional fishing. The town of Muisne is located at the northern end of the estuary on an offshore island with no vehicular access. Tourists and residents arrive and depart via a small car ferry that takes no more than a couple of minutes to cross the river. Tourism is popular in the summer months with several hotels on the island, although it is not a major tourist resort in the realm of the nearby Atacames. The beach is a high-quality sand beach with shallow waters, attracting tourists to Muisne in the dry months of June, July, and August as well as during the various Ecuadorian holidays. The summer months also see the arrival of whale-watching tourists who come to see the humpback whales breeding just offshore. Carnival is a particularly popular time when the hotels, guesthouses, and beaches are packed. The region is well connected to the rest of Ecuador, as it is located on the major north-south highway and is near the provincial capital and oil port of Esmeraldas.

The population of Muisne town was 5925 at the 2010 census with an equal number of people spread around the numerous small communities that dot the estuary. The 2008 population estimate within 5 km of the entire estuary is 13,977, and the population estimate within 10 km of the entire estuary is 23,168 (Oak Ridge National Laboratory and UT-Battelle LLC 2014; Hamilton 2012). Due to the various employment opportunities in the region, especially tourism, only approximately one-third of residents are reliant on the mangrove forests for their livelihoods. As such, 18% of the residents of this estuary own a boat, whereas only 16% own a vehicle and 35% of the estuarine residents own fishing tackle (Collins 2010). Education and income appear to have a strong relationship with both a household's and an individual's reliance on the mangrove forest economy in and around Muisne (Collins 2010).

As opposed to Cayapas-Mataje, in which almost the entire economy is based around the mangrove forests, in Muisne, it appears that the mangrove forest economy is still important, but it is not the sole livelihood provider. It must be remembered though that this estuary has been stripped of almost all of its mangrove forest, so it is likely that mangrove forest reliance was much higher historically, though it remains important today. In earlier periods the estuary was a rich cockling ground with residents able to harvest as many as 400 concha negra per day in the early 1990s, but this rapidly declined by the later 1990s (MacKenzie Jr. 2001). Approximately half the concha negra harvesters that remain today report loss of their historic collection areas due to the expansion of shrimp farms in the area (Rosero Moya and Santillan Salas 2011). As opposed to other estuaries where foreign companies own and operate shrimp farms, local residents appear to own, manage and work a large proportion of the shrimp farms of this region.

3.3.3 Biodiversity

The mangrove forests of Muisne are known to be an important sea turtle nesting site, providing a safe repository for the eggs of the hawksbill turtle, the green sea turtle, the leatherback turtle and the olive/Ridley sea turtle (also called Pacific Ridley sea turtle, both *Lepidochelys olivacea*) (Ministerio del Ambiente 2018). These species are currently listed as critically endangered (population decreasing), endangered (population decreasing), vulnerable (population decreasing), and vulnerable (population decreasing) respectively (IUCN 2010). Indeed, the hawksbill turtle is the only surviving member of its genus *Eretmochelys*. Muisne is one of one a few known sites along the Pacific Coast of Ecuador were the olive/Ridley sea turtle is known to nest (Ministerio del Ambiente 2018).

Over 90 fish species are found in the estuary including Ecuadorian estuarine regulars such mullet, snapper, and sargeant fish, but occasionally larger deep-water fishes such as dorado can be observed as this region's rocky outcrops support the locally-consumed corvina (Ministerio del Ambiente 2018). In addition to the fish are octopuses, bivalves, and crustaceans. Mammals include a large population of howler monkeys and numerous river otters. Humpback whales breed at the northern end of the study area within sight of the shoreline during the dryer and more touristic months of June, July, and August. Birds include frigate birds, kingfishers, large osprey, and a variety of parrots.

3.3.4 *Protected Status*

As opposed to REMACAN which is a large singular site, the Manglares Estuario del Río Muisne Wildlife Refuge Vida Silvestre Manglares of the Rio Muisne Estuary Refuge is essentially a group of 25 discrete mangrove stands comprising approximately 3173 ha (Ministerio del Ambiente 2018). The region has been heavily impacted by mangrove deforestation, and just as with REMACAN, the protected status was granted to preserve the mangrove forest and traditional mangrove forest economy that was at risk of disappearing (Ministerio del Ambiente 2018). The wildlife refuge status does not offer the same level of protection as a state-owned National Park such as REMACAN. The wildlife refuge land can remain in private ownership but is protected by decree and agreements with landholders the Ministry of the Environment (MAE). A small section of mangrove forest was historically protected by a private group and is known as The Mangroves of the Muisne River Estuary Wildlife Reserve. No more than a couple of hectares in size, it was created to protect a dwarf mangrove species located in the estuary. This private and voluntarily protection regime appears to be no longer active since the 2017 decree protecting all the remaining mangrove forests in the country. The not-for-profit advocacy group Fundación de Defensa Ecológica (FUNDECOL) is active in the estuary, managing the resource alongside the MAE (Herrera and Elao 2007) and acting as a liaison between residents and the government (Collins 2010).

Slightly offshore of this region is the Galera San Francisco Marine Reserve. According to the World Database of Protected Areas and the MAE, Galera San Francisco Marine Reserve is just north of the study area and continues out to the continental shelf at a water depth of 800 m. (Ministerio del Ambiente 2018; IUCN and UNEP 2016). Galera San Francisco Marine Reserve was Ecuador's first continental marine reserve and was established to protect the remarkably high native biodiversity found in these waters, with marine biodiversity levels close to that of the infamous Galapagos Islands farther offshore (Ministerio del Ambiente 2018). Unlike the mangrove forest area, this area to the north has coral reefs as well as rocky coastlines and cliffs.

3.4 Cojimies Estuary and the Vida Silvestre Manglares of the Rio Muisne Estuary Refuge: The Loss of a Functioning Estuary

Figure 3.6 depicts the study area (Fig. 3.1, iii), referred to in this manuscript as Cojimies Estuary, which is split almost equally between Esmeraldas Province and Manabí Province. It is likely the least studied estuary in Ecuador (Herrera and Elao 2007), and hence much of the factual information comes from the author's visits to the estuary, community and individual surveys by Collins (2010) as part of a

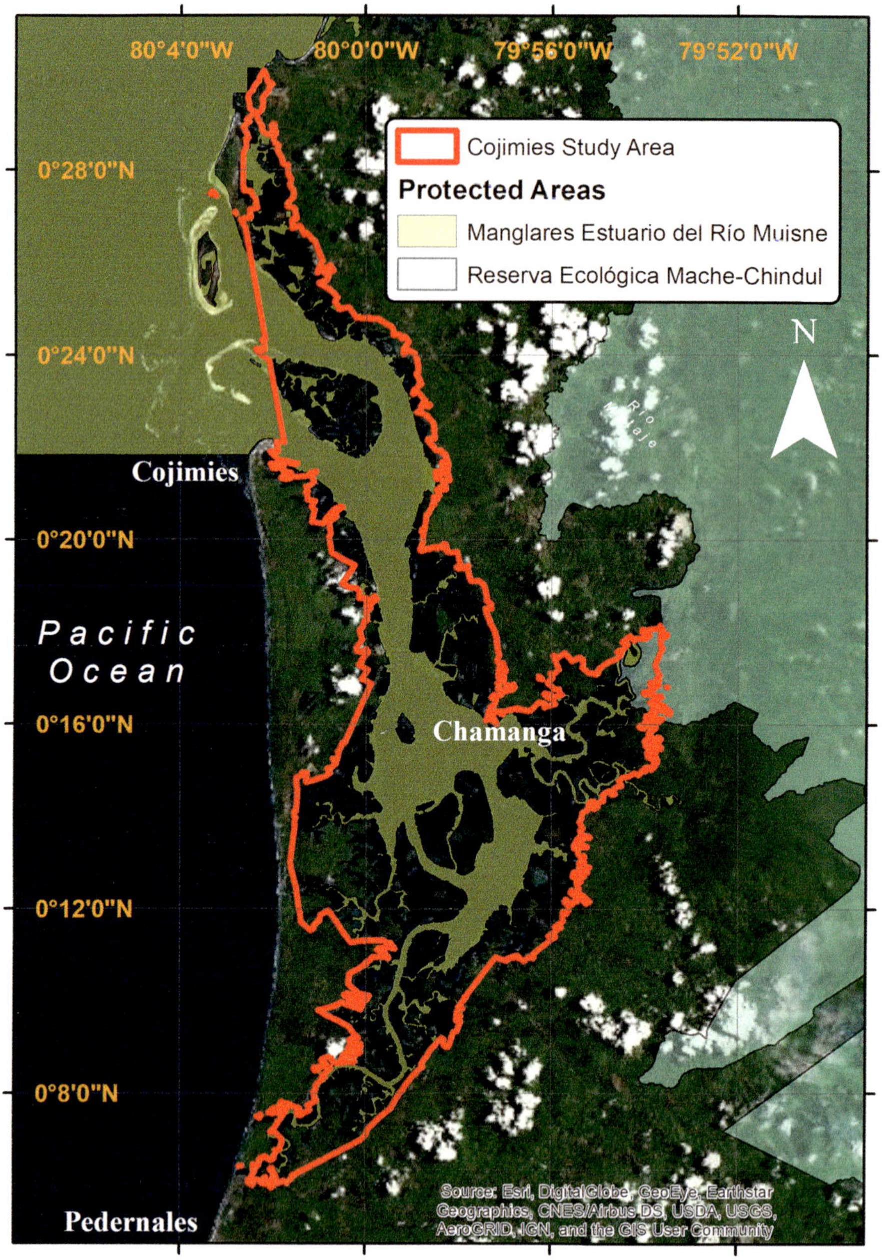

Fig. 3.6 Cojimies estuary study area

masters-level degree, and the grey literature including reports compiled by aid agencies.

Numerous small traditional villages dot the estuary. These include the communities of Zapotal, Daule, Mocoral, San José de Chamanga, Beche, and Cheve with traditional fishing and agriculture as the dominant occupations. The study area is 35,435 ha. A few kilometers past the southern end of the estuary is the regionally important city of Pedernales which was at the epicenter of a major magnitude 7.8 earthquake in 2016 that destroyed or damaged almost all the structures in the city and most homes in and around the study site. Cojimies itself is a small artisanal fishing community with sand roads, small wooden homes, and a pristine beach located at the mouth of the estuary on the ocean side.

The estuary contains three major islands, Isla de Cojimies, Isla el Moro de Juananu, and Isla Angardonie which is divided into two at high tide by a small creek (Fig. 3.6). At least a dozen or so small rivers flow into the estuary along with the major rivers of Rio Cojimies and Rio Cuasa which from the east (Fig. 3.6). Of the 12 major rivers, at least eight are now mostly dry (Crawford 2010). Only a small peninsula separates the northern tip of Cojimies Estuary and the southern tip of Muisne Estuary, although they are distinct. The few remaining mangrove stands in the Cojimies Estuary are in and around Moracumbo, Isla Esmeraldas, Isla Jupiter, Isla El Morro, Guadual, Bolivar, Pedro Carbo, Duale, Salima, Chamanga, Cojimies, Guadual, and El Churro (Ministerio del Ambiente 2018). The major remaining mangrove stands in this area were given protected status in 2003 as part of the Refugio de Vida Silvestre Manglares Estuario del Rio Muisne (Collins 2010), and this protection was enhanced in 2017 (Ministerio del Ambiente 2018). The species composition is 70% *Rhizophora mangle*, 5% *Avicennia germinans*, and 25% *Laguncularia racemosa* (Hamilton et al. 2017).

3.4.1 Climatic & Environmental Conditions

The Cojimies Estuary likely is somewhat influenced by the cold Humboldt/Peruvian current (Chap. 1, Fig. 1.4), but the few remaining mangrove forests are in a secluded estuary that is approximately 30 km from the mouth of the estuary, with the estuary having a narrow oceanic opening compared to its large size. During early February the water temperature is typically at its highest with a maximum temperature of 29 °C. During early August the water temperature is coldest with a minimum temperature of 23 °C. Annual rainfall averages 922 mm but with much higher levels during El Niño years and much lower levels during La Niña years. Cojimies Estuary has a pronounced and long dry season from July to November (Fick and Hijmans 2017). The average monthly air temperatures show limited seasonal variation with April being the hottest month with an average daily temperature of 26.1 °C, and September being the coolest month with an average daily temperature of 24.1 °C. The Köppen-Geiger climate classification is Aw, or Equatorial and winter-dry (Kottek et al. 2006).

3.4.2 Socioeconomic Conditions & Livelihoods

Cojimies estuary is an ethnically mixed region with Afro-Ecuadorians in the minority. The estuary is dotted with numerous small fishing villages. The residents of the villages surrounding the Cojimies Estuary, particularly those towards its interior, have traditionally made their income from the goods and services offered by the mangrove forests. However, today, the mangrove forest is almost entirely cleared, and incomes, as well as livelihoods, have declined along with the mangrove forest. In response to these challenges, income and livelihood recovery efforts supported historically by the United States Agency for International Development (USAID) aim to restore the estuary so that local inhabitants can again achieve sustainable livelihoods (Crawford 2010; Herrera and Elao 2007). Unfortunately, these USAID activities were suspended for a large part of the late 2000s and beyond, as USAID could not undertake projects in Ecuador.

The 2008 population estimate within 5 km of the estuary is 12,812, and the population estimate within 10 km of the estuary is 18,232 (Oak Ridge National Laboratory and UT-Battelle LLC 2014; Hamilton 2012). Unlike Muisne, with a strong tourist industry, or REMACAN with a strong mangrove forest economy, Cojimies appears to have only modest agriculture as an alternative to the now depleted mangrove forest economy. Infrastructure is limited, and until the last few years, the only passable road to Cojimies was to drive the sand beach at low tide, although a new coastal road built in 2014 now parallels the beach. Infrastructure in the small village is limited with no sewage service, potable water supply, garbage collection, or other essential services (Herrera and Elao 2007). Of all the coastal Ecuadorian communities surveyed, in the Cojimies Estuary study area, only 10% of residents pay for electricity, which is by far the lowest proportion with the next lowest being 30%. (Collins 2010). The towns at the interior of the estuary appear to offer few livelihood opportunities and are economically challenged with people living in poverty. As with the other northern estuaries, the collection of shrimp and mollusks are often conducted by women in Cojimies Estuary. As of the early 2000s though, shellfish harvesting within the estuary has been suspended as both the blue crab and brown crab are overexploited (Herrera and Elao 2007) and species like the bivalves have entirely disappeared. Few alternates livelihood options appear to exist. Even the important estuarine fisheries are so heavily depleted in Cojimies Estuary that they are of little economic value, and this has caused social unrest among the fishers of the region.

3.4.3 Biodiversity

Biodiversity is threatened by the high levels of poverty in the region (Herrera and Elao 2007). All the critical fisheries are threatened with many such as the crab and concha fisheries almost locally extinct. Other threats to biodiversity include sedimentation that has damaged the estuarine ecosystem and the removal of water for agriculture before it reaches the estuary, often causing the input rivers to run dry (Herrera and Elao 2007). No comprehensive inventory of species or species counts could be obtained for Cojimies Estuary, but it is likely a more degraded biodiversity situation than the adjacent Muisne Estuary.

3.4.4 Protected Status

According to the World Database on Protected Areas (WDPA) and the MAE, the few remaining mangrove stands are protected under the same decree that created Vida Silvestre Manglares of the Rio Muisne Estuary Refuge (IUCN and UNEP 2016; Ministerio del Ambiente 2018). This patchwork protected status is fundamentally a delineation of the few remaining mangroves stands. The protected area is managed by the MAE with input from FUNDECOL, although it is noted that FUNDECOL representation declines further into the estuary and representation is entirely lacking in the southern interior of the estuary (Collins 2010). Indeed the majority of residents in and around Cojimies Estuary report that they are not aware of any entity that monitors the use or destruction of the mangrove forests (Collins 2010). Adjacent to the estuary and running parallel to the western border of the study area is the Chindul Ecological Reserve which is a sizeable terrestrial reserve that extends into the coastal mountains.

3.5 Chone Estuary and Vida Silvestre Islas Corazón y Fragatas Wildlife Refuge: The Development of an Eco-City

Figure 3.7 depicts the study area (Fig. 3.1, iv), referred to in this manuscript as Chone Estuary, this estuary is located between the dammed portion of the Rio Chone in the east and the mouth of the Pacific Ocean in the west. Chone Estuary is the most southern of Ecuador's northern estuaries and is centrally located in Manabí Province. As opposed to all the study sites further north, Chone Estuary is surrounded by cliffs on both its northern and southern banks. These steep slopes combined with the dam in the interior of the estuary make estuary delineation relatively simple. The only major river entering the estuary is the Rio Chone to the east (Fig. 3.7), and a dam entirely controls the river flow. East of the dam is the large Segua Wetland RAMSAR site, which was known to be densely covered with mangrove forest within the last hundred years, although none remain as the environmental conditions are no longer suitable for mangrove forests. Indeed, the Segua Wetland is no longer an estuary but a purely freshwater environment.

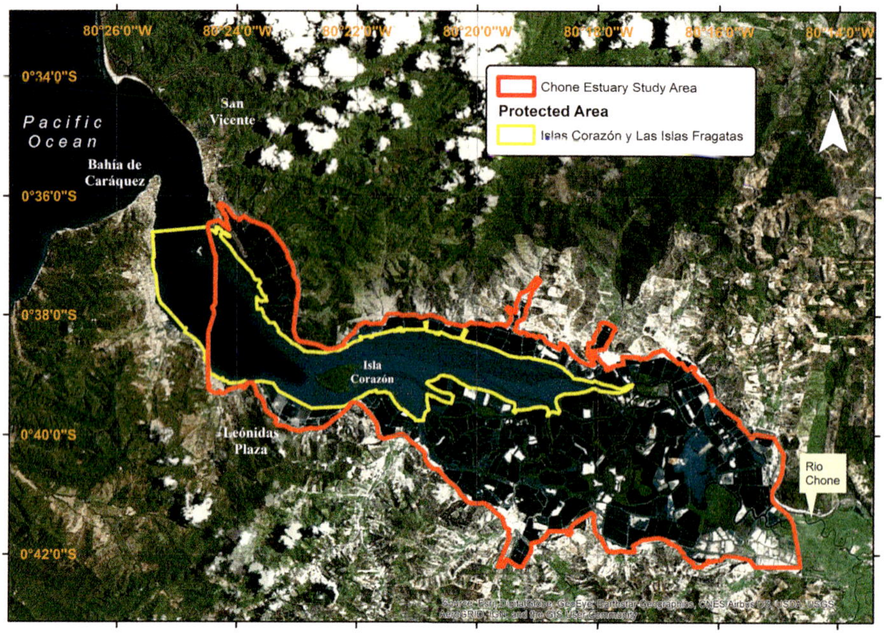

Fig. 3.7 Chone estuary and Islas Corazón y Fragatas wildlife refuge

The study area is 8655 ha of which 2812 ha is protected. (Fig. 3.7). Although the elevation in the study area is no more than 3 m, along the banks of both the northern and southern side of the estuary are highly erodible sand and marine sediment cliffs as high as 100 m. The interior of this estuary has extremely large and tall *Rhizophora mangle* stands similar to REMACAN, although they are far less numerous (Hamilton et al. 2017). Shrimp farms are now the dominant landcover within the estuary, although mangrove forest once dominated (CLIRSEN 2007). *Rhizophora mangle* constitutes the majority of the mangrove forest species composition with 75% of all the mangrove forest area, *Laguncularia racemosa* covers 15%, and *Avicennia germinans* accounts for the final 10% (Hamilton et al. 2017). The single most significant mangrove forest area remaining in Chone Estuary is Isla Corazón, and this has grown larger in recent years due to aggressive replanting. The 2008 population estimate within 5 km of the estuary was 37,783, and the population estimate within 10 km of the study site was 68,737 (Oak Ridge National Laboratory and UT-Battelle LLC 2008).

3.5.1 Climatic & Environmental Conditions

The Köppen-Geiger climate classification for the entire estuary is Aw, or Equatorial and winter-dry (Kottek et al. 2006), with a distinct dry period beginning around May and ending in November. Of all the study sites, Chone Estuary is most heavily impacted by the Southern Oscillation with extreme rainfall in El Niño events but limited rainfall during La Niña periods. The Southern Oscillation can also drastically alter nearshore water temperature. The average annual rainfall in the estuary is reported as a little over 500 mm, and the average temperature is reported as just under 25 °C (Bodero 1993). During an El Niño year, the rainfall can be as much as ten times higher than during a La Niña year. During the dry season the Rio Chone, the only real river feeding the estuary, has almost no detectable flow; this condition also occurs regularly during La Niña periods (Stram et al. 2005).

Chone Estuary is an aging estuary, likely in the process of disappearing from natural causes (Odum et al. 1991; Arriaga et al. 1999). The loss of terrestrial and aquatic forests, the earthquake-driven sedimentation events, and the El Niño-driven sedimentation events likely hasten the estuary's disappearance. It is apparent that sedimentation is overwhelming the estuary as the highly erodible sand and marine deposits continue to find their way into the estuary where the water flow rates are not high enough to subsequently remove the sediment.

3.5.2 Socioeconomic Conditions & Livelihoods

The city of Bahía de Caráquez at the mouth of the estuary is a tourist center with numerous condominiums, hotels, and guesthouses catering to international expatriates and visitors from the highlands. The tourism is mostly limited to Bahía de Caráquez with tourism occurring mostly in the dry season as well as over holidays including Carnival. The city is partly on the estuary and partly on the ocean. Aside from the fish, shrimp, and other seafood served in the tourist restaurants, the city of Bahía de Caráquez has only limited livelihood connections to the estuary or mangrove forests. Tourists in Bahía may visit the mangrove forests of the estuary for a day trip, and some even venture inland as far as the Segua Wetland.

Away from the major cities, numerous small fishing communities reside on the terrestrial edge of the estuary, and residents in these communities have traditionally made their living from the goods and services offered by the estuary. For example, on the northern shore of the estuary in the rural communities east of the town of San Vicente (Fig. 3.7), 80% of the households are dependent on the estuary for their income from fishing, while only 10% are reliant on either tourism or agriculture. In Sucre community, 80% of the households rely fishing and 20% on the collection of mollusks and crustaceans (Capitulo 2008). Capitulo (2008) provides an exhaustive and detailed list of livelihoods provided by the mangrove forests, including building material, charcoal, tannins, alcohol, direct foodstuffs, and medicinal uses.

The artisanal fishers appear well organized in this region, representing themselves through a union (Fig. 3.8, Panel 1) as opposed to relying on non-governmental organizations (NGOs) as in Muisne or Cojimies. Many of the artisanal fishers now rely on ecotourism as opposed to fishing or harvesting in the estuary. For example, the small fishing community in and around Portovelo on the north side of the estuary has rebranded itself as the Isla Corazón Ecotourism Community (Fig. 3.8, Panel

Fig. 3.8 Chone estuary and ecotourism opportunities
Caption – Panel 1 is the artisanal fisher union building in Bahia de Caraquez. Panel 2 is taken from the cliffs around the estuary and shows the Isla Corazón mangrove forest in the top left and shrimp farms in the top right. Panel 3 is an artisanal community canoe tour through the mangrove forests of Isla Corazón. Panel 4 is a former fishing community that has rebranded itself as an eco-tourism community. Panel 5 is a sign carrying the eco-city label in the bottom left and revealing the secrets of Isla Corazón. Panel 6 shows abandoned fish and shrimp processing plants.

4) and offers tourists shrimp lunches and tours of the estuary. Particularly popular is the canoe track through the center of Isla Corazón (Fig. 3.8, Panel 2 & 3). These ecotourism-driven former fishermen are responsible for much of the mangrove forest replanting that occurs in the estuary. Fishing and crab or concha harvesting are

much reduced from earlier periods, and the region is dotted with former processing plants for shrimp and other species that are no longer in operation (Fig. 3.8, Panel 6). It is likely these plants also served the nascent shrimp farm industry in the 1980s and 1990s before all the processing moved to the port area of Guayaquil.

The region around Chone Estuary has suffered recent catastrophic El Niño rainfall events. As a result, the estuary has sedimentation issues that impact water quality (Hamilton et al. 2015; Leon et al. 2016), leading to a marked decrease of local livelihood options within the estuary (Ubilla 1998). Indeed, the environmental degradation caused to the estuary by mangrove deforestation and increased sedimentation inputs was valued at over $45 million in the mid-2000s (Capitulo 2008).

3.5.3 *Biodiversity*

The Chone Estuary is best known as a globally important nesting site for the frigatebird, for which the refuge is named, with large colonies found in and around Islas Corazón. Twenty-nine bird species are confirmed, but the refuge and community guides report as many as 90 bird species, other species include 22 crustaceans, 49 mollusks, and five mammals (Capitulo 2008). The MAE reports more than 100 birds species in the estuary (Ministerio del Ambiente 2018). An endangered reptile and the vulnerable boa snake are additionally found in the estuary (Capitulo 2008). The MAE lists the important species in the estuary as snails, shrimps, crabs, sawfish, croaker, snapper, snook, and mullet (Ministerio del Ambiente 2018).

3.5.4 *Protected Status*

Since 2002, 2812 ha or one-third of the study area is protected as part of the Vida Silvestre Islas Corazón y Fragatas Wildlife Refuge (Fig. 3.7). The MAE manages the resource with the express intent of preserving the remaining mangrove forests and the livelihoods that they support. Before being a protected site, the Chone Estuary area was subject to the decade-long government and community partner Special Area Management Plan (ZEM). The goal of the plan was to manage land use and activity around the estuary and provide local stakeholders with incentives to participate in the recovery of the estuary (Coello et al. 1993). The Swedish branch of the Nature Conservancy, numerous international NGOs, and a variety of local environmental actors actively work to return the estuary to its original condition. The local fishing syndicate on the northern banks of the estuary actively replant mangroves around Isla Corazón, and local and international students can regularly be found in the estuary working on environmental restoration projects. Numerous murals and signs around the region point to the role of the environment in the region as well as highlighting the importance of the mangrove forests and estuarine environment (Fig. 3.1, Panel 4).

More important than the protected status is the change in attitudes among the region's residents towards mangrove forest and other conservation efforts. In response to three major disruptive events around the late 1990s, Bahía de Caráquez declared itself an eco-city (Fig. 3.8, Panel 5) and began to take numerous steps to improve the surrounding environment with an emphasis on reforestation, protection of the estuary, and ecotourism. The first event was the major1997 El Niño period that resulted in heavy rains leading to numerous landslides causing the loss of life and severe erosion. Much of the blame for the landsides was placed on deforestation. These El Niño-induced landslides resulted in a massive amount of sediment entering the estuary as entire mountainsides, consisting of sand and marine sediment, collapsed into the estuary. The next incident that followed the El Niño event was a magnitude 7.1 earthquake in 1998 which also caused major landslides into the estuary and loss of life (Fernandez and Yepes 1998). The final incident was a widespread outbreak of white spot shrimp disease in the shrimp farms in the late 1990s that decimated the estuarine shrimp industry.

3.6 Isla Puná North and Isla Puná

Figure 3.9 represents the study sites of Isla Puná North (Fig. 3.1, v) and Isla Puná (Fig. 3.1, vi). These are two distinct estuaries and are therefore two different study areas. The island is by far the largest in the Gulf of Guayaquil and sites near the mouth of the Rio Guayas. The Gulf of Guayaquil, known as the outer estuary, starts at the northern end of the island and incorporates the Morro Channel, the Jambeli Channel, and extends out into the Pacific Ocean. The island is approximately 40 km from north to south and approximately 30 km from east to west along its longest and widest axis and is approximately 850 km^2 in size. The entire island is a parish in the Guayaquil canton within Guayas Province. The island is separated from Guayas by the 4 km wide Morro Channel in the west and by the 15 km wide Jambeli Channel in the east (Fig. 3.9). The island is dotted with small traditional communities that often lack regular power supply or running potable water. Those with a population over 100 are Agua Piedra, Bellavista, Campo Alegre, Estero de Boca, Puná Vieja, Subida Alta, and Zapote (Madsen et al. 2001). No paved roads exist on the island and boats are the primary mode of transportation. The only way on and off the island is by boat. The total population of the island was 6498 at the 2000 census and 6769 at the 2010 census (Brinkhoff 2011).

The island is lined with small lagoons, mudflats, saltmarshes, and mangrove forests (Healy et al. 2002). The northern mangrove forests are located along the shoreline and at the mouth of a handful of small and often intermittent rivers. The larger mangrove forest in the southwest is an estuarine system where the mangrove forests extend as much as 15 km inland from the Gulf of Guayaquil and the Jambeli Channel (Fig. 3.9). The northern study area is 7154 ha in size, and the southeastern study area is 31,245 ha and would cover as much as one-quarter of the entire island if it was a terrestrial landcover.

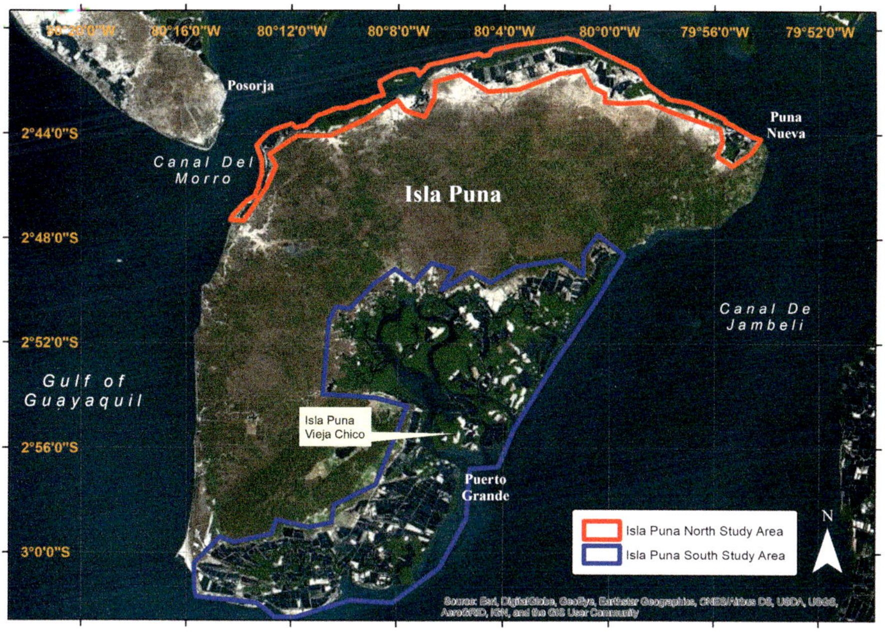

Fig. 3.9 Isla Puná north study area and Isla Puná study area

3.6.1 *Climatic & Environmental Conditions*

The island is in the Köppen-Geiger climate category Aw, known as Equatorial and winter-dry. (Kottek et al. 2006). The island's climate is heavily influenced by the cold Humboldt/Peruvian Current meaning rainfall is limited on the island with the dry summer period lasting between 8 and 10 months with May to December experiencing little rainfall but also having very few clear days. Again, rainfall variation is controlled by the Southern Oscillation, with rainfall as high as 3929 in El Niño years. The nearest rain gage to Isla Puná is on the mainland in Playas about 25 km away and indicates average rainfall of about 390 mm in this area (Madsen et al. 2001). The air temperature generally is at its lowest in late July and early August, with daytime averages around 24.5 °C, and its highest in late March and early April, with average high temperatures a little over 27.1 °C. Water temperature is again determined by the Southern Oscillation, with El Niño years resulting in noticeably warmer water temperatures. Outside of Southern Oscillation-driven extremes, the water temperature varies between a low of 24.5 °C in the middle of the dry period in September and a high of 26 °C during the wet winter months. This measure is taken from water in the Jambeli Channel (Fig. 3.9). The tidal range is 1.8 m, more limited than that of Guayaquil (Twilley et al. 2001).

3.6.2 Socioeconomic Conditions & Livelihoods

Without the regular MAE surveys that go along with protected status, assessing the modern livelihood situation of the Isla Puná communities is difficult. Madsen et al. (2001) note that the younger generation is drawn to shrimp farming or cash labor in Guayaquil and traditional occupations such as agriculture are becoming less common, although fishing remained the principal occupation on the island as of 2001. It is estimated that approximately 90 cockle fishers lived on the islands in the early 2000s (MacKenzie Jr. 2001). The concha harvest does not appear to be woman-dominated in this estuary. Historically though, concha negra harvesting was necessary; in the late 1990s, as much as 80% of the population of communities such as Campo Alegre harvested mollusks (MacKenzie Jr. 2001).

Today, it appears that shrimp farming is the largest commercial industry on the island and 45% of people are involved in the artisanal wild fishery (Elizaldes Consultora 2015). The entire island lacks essential services such as fresh water, sewer, consistent and stable power, and medical facilities (Elizaldes Consultora 2015), and the rates of poverty are as high as 90% (Collins 2010). About half of the population are without power and those that have power report that outages of up to 10 h a day are common (Collins 2010). Indeed, Isla Puná was recognized as having the highest levels of poverty of all parishes during planning for Refugio de Vida Silvestre Manglares el Morro site located nearby on the Guayas mainland (Ministerio del Ambiente del Ecuador 2010).

3.6.3 Biodiversity

Five species of mangroves are documented on Isla Puná. These species are *Avicennia germinans*, *Conocarpus erectus*, *Laguncularia racemosa*, *Rhizophora mangle*, and *Sarcocornia fruticosa* (which is now considered a mangrove associate); *Rhizophora mangle* constitutes 95% of all mangrove forests on the island (Madsen et al. 2001). Madsen et al. (2001) provide an exhaustive list of 100 flora species on the island. The island is also known among ornithologists for its richness in bird species that include gulls, pink herons, white herons, pelicans, albatross, cucube, blue-footed boobies, cormorants, frigatebirds, sea swallows, and zarapitos (Elizaldes Consultora 2015). Impact fauna includes the Jaguarundi wildcat (*Herpailurus yagouaroundi*) and most importantly for tourism, the Guayaquil common bottlenose dolphin (*Tursiops truncates*) (Elizaldes Consultora 2015) that frequent the Morro Channel (Fig. 3.9). Twenty-six species of commercially important fish are listed as inhabiting the nearshore waters, as well as four crustaceans and four bivalves (Elizaldes Consultora 2015).

3.6.4 *Protected Status*

Numerous agencies report that Isla Puná has been protected since 2009, although the legal standing of the designation remains unclear. The MAE does not list it as a protected area, and it is omitted from the current SNAP database. What is known is that the remaining mangrove forests fall under a sustainable use and custody agreement administered by the MAE in place since the late 1990s or early 2000s, and much of the remaining terrestrial forest is under a conservation and sustainable use agreement with the Socio Bosque Program (Elizaldes Consultora 2015; Collins 2010).

3.7 Guayas Estuary and a Patchwork of Protected Areas: Mangrove Forests in a Dense Urban Environment

Figure 3.10 depicts the study of Guayas Estuary (Fig. 3.1, vii), the patchwork of protected status in and around the estuary, and the city of Guayaquil in the north of the estuary. According to current 2018 census estimates, Guayaquil is the largest city in Ecuador with a population of 2,350,915. As the entire population of Ecuador is estimated at 17 million, Guayaquil contains approximately 13% of Ecuador's

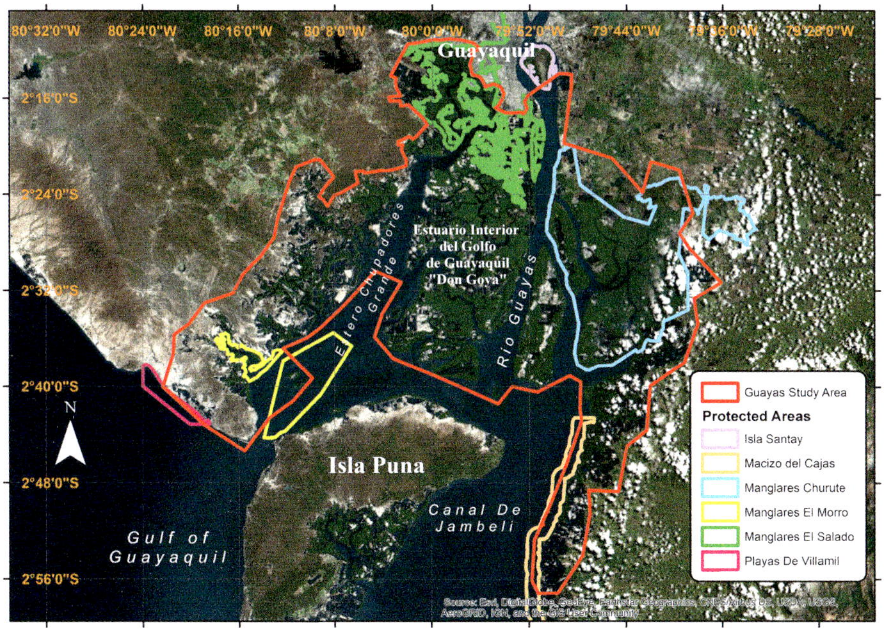

Fig. 3.10 Guayas estuary and a patchwork of protected areas

entire population. The Guayas Estuary study area is a classic flat large wetland area with the significant inflow coming from the Rio Guayas which itself drains a large portion of the Western Cordillera of Ecuador including much of the highlands. The Rio Guayas itself (Fig. 3.10) is a confluence of two other major rivers, the Rio Babahoya and Rio Daule, that merge at the northern end of the city. The Guayas Estuary, often referred to as the inner estuary, extends approximately 75 km from Isla Puná to the maximum tidal extent of the Rio Guayas. The Rio Guayas has a flow rate of 20 km^3 $year^{-1}$ (Stevenson 1981). Other major input rivers include Rio Taura, Rio Churute, Rio Arenilles, and Rio Jubones, but all of these freshwater inputs combined are far less then what is provided by the Rio Guayas.

The major mangrove forests are in the intertidal wetland between the Rio Guayas and the Grande Chupadores Estuary south of the city and on the eastern banks of Guayas Estuary, following the Jambeli Channel and continuing to the land border with El Oro Province (Fig. 3.10). Some more limited mangrove forests exist on the western edge of the Grande Chupadores Estuary. The five mangrove species present are *Rhizophora harrisonii*, *Rhizophora mangle*, *Avicennia germinans*, *Laguncularia racemosa*, and *Conocarpus erectus*. The dominant mangrove species is *Rhizophora mangle*, although Twilley et al. (2001) report the dominant mangrove as *Rhizophora harrisonii*. *Rhizophora harrisonii* does dominate closer to Guayaquil but much less so in the areas further south, and this dominance in the more accessible areas is likely the cause of the confusion over the dominant species. In 1969, the mangrove forest within the estuary covered 122,615 ha and constituted 61% of all Ecuadorian mangrove forest cover (CLIRSEN 2007). The interstate highway bridge over the Rio Guayas that runs from the city of Guayaquil to the city of Durán (−2.1585131°, −79.8549578°) is a good proxy for the northern extent of substantial mangrove forests.

Around the urban edges of Guayaquil, particularly in the south, high-density residential communities abut the mangrove forests with only a small channel separating them (Fig. 3.11). Indeed, commuter roads cross the mangrove forests on stilts between different neighborhoods in the city. Although some urban edges abut mangrove forest in El Oro province, Guayas Estuary is the only region where high-density large-city urban environments abut vast mangrove forests. The mangrove forests of Guayas Estuary that extend into the city are a federally protected site called Manglares el Morro, and the residential areas cannot expand into them without a change in federal legislation (Fig. 3.10). Substantial low-lying dense urban areas are shown in Fig. 3.11 and beyond within the city of Guayaquil were once likely all mangrove forests themselves, and indeed this is depicted in colonial maps such as that produced Minget (1741).

3.7.1 Climatic & Environmental Conditions

Guayas Estuary resides in the Köppen-Geiger climate category Aw, known as Equatorial and winter-dry (Kottek et al. 2006). The Guayas basin average rainfall level is 885 mm per year, but like other study sites, the range varies from a low of

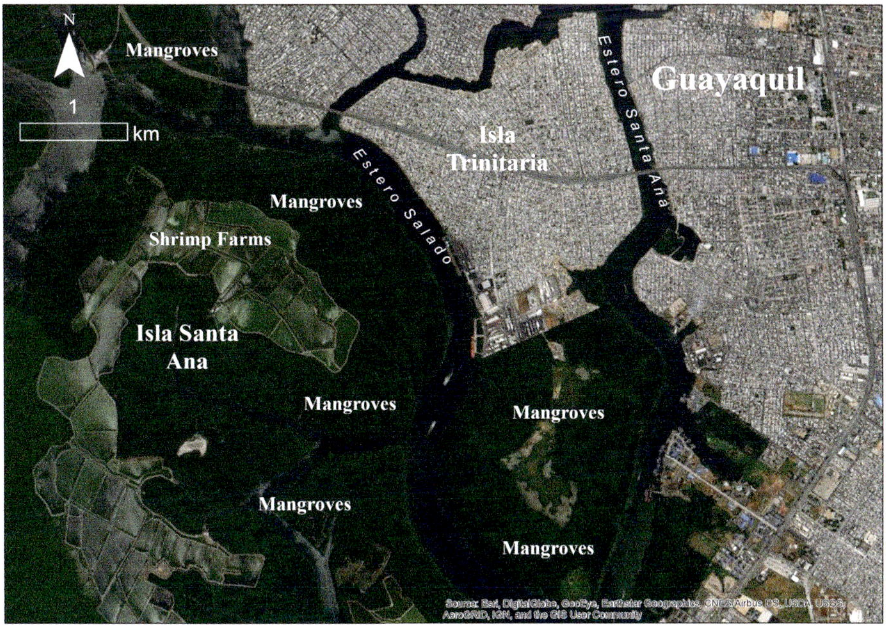

Fig. 3.11 Mangrove forests in an urban environment

only 400 mm per year and a high of 3800 mm per year when influenced by either La Niña or El Niño respectively (Cucalón 1983, 1989). Almost all of this rainfall occurs between December and May. The tidal range is between 4.5 and 4.6 m (NOAA Center for Operational Oceanographic Products and Services 2018) at the city of Guayaquil. The annual mean air temperature shows little variation and is between 24 and 27 °C (Cucalón 1983) with a preponderance of overcast days even during the dry season. During the dry season, the water temperature varies between 21.5 °C at the southern end of the estuary to 25 °C further north in the estuary (Stevenson 1981) and increase by about 3 °C during the wet season (Twilley et al. 2001). Across each of the major Guayas Estuary regions, the water temperatures are listed as between 22 and 30 °C, with a mean of 27 °C in and around the mouth of the Rio Guayas, between 24 and 31 °C, with a mean of 27 °C in and around the mouth of the Rio Salado, and between 23 and 30 °C, with a mean of 25 °C in and around the mouth of the Rio Churute (Twilley et al. 2001).

3.7.2 *Socioeconomic Conditions & Livelihoods*

Obviously, in a city of 2.5 million, the livelihoods in Guayaquil are varied, and the vast majority of the population's livelihoods are not interconnected in a meaningful manner to the goods and services offered by the estuary. On the opposite side of the

Rio Guayas from Guayaquil is Ecuador's sixth largest city of Durán. Like Guayaquil, livelihoods in Durán are also varied and mostly independent of the estuary. Outside of the major cities and their suburbs in other parts of Guayas Estuary, this is not the case. Isolated and artisanal fishing communities live in similar circumstances to the other study areas. For example, the populations in and around Manglares Churute Ecological Reserve experienced a complete lack of adequate housing, sanitation or medical facilities well into the 1990s and an illiteracy rate of around 30% (Yanez 1996). These population's primary livelihood is the artisanal fishery and shrimp farming (Yanez 1996).

On the other side of the Guayas Estuary in and around Manglares el Morro, the three analyzed communities have poverty rates of 94%, 94%, and 95% with extreme rates of poverty as high as 56%, 56%, and 60% (Ministerio del Ambiente del Ecuador 2010). Again, the major occupations in Manglares el Morro are artisanal fishing and tourism. As of the late 1990s, the small community of Sabana Grande near Manglares el Morro had 35 cockle collectors (MacKenzie Jr. 2001). Even at the northern end of the estuary in closer proximity to Guayaquil, 102 of the 173 socio-economic actors were artisanal-style fishers who were active along the reserve as late as 2006 (Ministerio del Ambiente 2018). The presence of small-scale artisanal aquaculture is important around the Estuario Interior del Golfo de Guayaquil – Don Goya Ramsar site in the center of Guayas Estuary (Fig. 3.10) where fishers manage small ponds with only one to two shrimp cycles per year, manage a long-term sustainable crab fishery, and harvest numerous mollusks (Maldonado et al. 2014).

3.7.3 Biodiversity

Manglares Churute Ecological Reserve alone boasts over 300 bird species, a host of non-fish marine life that includes crocodiles, alligators, red crabs, concha prieta, mussels and oysters, as well as numerous mammals including bats, opossums, raccoons, jaguars, ocelots, howler monkeys, and wild boars (Ministerio del Ambiente 2018). At the northern end of the estuary in Manglares el Salado, are listed red crabs, oysters, mussels, shrimp, catfish, bocachicos, and bass; and among the mammals are listed the crab-eating fox, river otters, the two-toed sloth, and anteaters (Ministerio del Ambiente 2018). On the eastern side of the estuary in and around Manglares el Morro, 80 species of birds are documented including blue-footed boobies, frigatebirds, and pelicans. The most critical mammal from a tourist perspective is the consistent presence of the common dolphins that appear to reside on this side of the estuary year-round.

The artisanal fishers of Manglares el Morro are known to catch mojarra as well as *Mugil cephalus* (lisa), *Galeichthys peruvianus* (catfish-lisa), *Catfish pinnimaculatus* (moustached catfish), *Oreochromis mossambicus* (tilapia), *Pomadasys schyrii* (snorer), *Centropomus spp.* (snook), *Cynoscion spp.* (corvina); the mussels and clams *Mytella strigata*, *Mytella guyanenss*, *Protothaca asperrima*, *Corbula inflata*, *Nucula colombiana*, *Macoma siliqua*, *Tellina spp.*, *Nucula colombiana*; the oysters

Ostrea iridescens and *Crassostrea columbiensis;* the swimming crabs *Callinectes arcuatus* and *Callinectes toxotes*; the crabs *Ucides spp.*, *Uca spp.*, and *Uca panamensis*; and of course, the shrimp, *Litopenaeus spp.*

3.7.4 Protected Status

Figure 3.10 depicts a patchwork of federally protected status in the Guayas Estuary. Within the study area are located all or portions of the MAE administered Manglares Churute Ecological Reserve, La Reserva de Producción Faunística Manglares el Salado, Área Nacional de Recreación Isla Santay y Gallo, Refugio de Vida Silvestre Manglares el Morro, Área Nacional de Recreación Playas de Villamil, and finally an isolated sliver of the much larger Macizo del Cajas Biosphere Reserve (Jiménez Tapia et al. 2013) is now located in Guayas Estuary. The combined mapped area of all parks (excluding any areas outside of Fig. 3.10) is 94,296 ha of which 69,229 ha falls inside of the Guayas Estuary Study Area (Fig. 3.10).

Área Nacional de Recreación Playas de Villamil is almost outside of the study area, but the boundary of the park boundary makes an intentional deviation to include a small mangrove forest in the community of Data de Posorja at the mouth of the Rio Data de Posorja. Aside from this singular mangrove forest, the park is aimed at providing recreation opportunities in a beach environment and contains no other mangrove forests. This portion of Macizo del Cajas Biosphere Reserve appears to be a marine preserve and is intentionally delineated to start at the outer and estuarine edge of the mangrove forest and extend out 2 km into the water, therefore it appears to contain no mangrove forest but is substantially inside the Guayas Estuary study area. Refugio de Vida Silvestre Manglares el Morro is a pure mangrove forest reserve that incorporates the two mangrove forest islands in the Morro Channel of Isla Manglarcito Grande, Isla Manglarcito Chico, and the remaining mangrove forest or Estuario el Ceibo on the mainland. The two islands appear to be formed from the accumulation of riverine sediments deposited from the surrounding rivers.

Only a small southern portion of Área Nacional de Recreación Isla Santay y Gallo is included in the analysis. This island, formed by the accumulation of sediments, has mangrove stands around its edge and along numerous inlets. The omission of the mangrove stands on the edge of this island is a delineation error that can no longer be rectified. This likely causes the data for Guayas to underestimate mangrove forest cover by under 0.25%. Isla Santay is Ramsar site 1041 and was incorporated in 2000 (RAMSAR 2006). La Reserva de Producción Faunística Manglares el Salado once again appears to be a dedicated mangrove forest reserve. It contains the remaining mangrove forests in the north of the estuary along the urban edge of the city of Guayaquil and extends far into the city along the estuaries, which contain some mangrove trees. Manglares el Salado delineation is a meandering polygon that essentially follows the outer edge of the remaining mangrove forests; that is, the park boundary was delineated from the remaining mangrove forest boundary. Manglares Churute Ecological Reserve is probably the most well-known of the

Guayas Estuary protected areas. Manglares Churute is Ramsar site 502 and was incorporated in 1990 (RAMSAR 2006). The Ramsar site and the MAE protected area delineated in Fig. 3.10 have slightly different boundaries but overlap enough to be considered a singular protected area.

Additionally, the major mangrove forests of the Guayas Estuary located south of Manglares el Salado between the Rio Guayas and Grande Chupadores Estuary are protected as Ramsar site number 2098 and was incorporated in 2013; this Ramsar site is known as Estuario Interior del Golfo de Guayaquil – Don Goyo. The Ramsar site is delineated in Maldonado et al. (2014).

3.8 The Mangrove Forests of El Oro Province and the Peruvian Border

Figure 3.12 depicts study area viii (Fig. 3.1) consisting of the entire coastal portion of El Oro Province and 15 km of Guayas province directly to the north of El Oro. This continuance is done to avoid subdividing a single homogenous estuary into two different study areas. The study area is 69,124 ha in size and runs approximately 75 km in a linear measure from northwest to southeast. The southern limit of the study area corresponds to the Peruvian border. The largest named estuary in this area is Grande Estuary (Fig. 3.12), which is bordered by Peru to the south and the

Fig. 3.12 The estuaries of El Oro province

urban center of Machala to the north. Grande Estuary is the largest estuary in El Oro province. The numerous estuaries of this region form countless islands of various sizes, many of which are sparsely inhabited.

Numerous fishing communities dot the estuary, the largest of which, Puerto Bolívar, has become mostly consumed as a Machala suburb and is now a sizeable commercial banana port. The community of Jambeli is located on the westernmost portion of the estuary and mostly acts as a tourist retreat to the residents of Machala (Fig. 3.1). The 2008 population estimate within 5 km of the estuary is 282,240, and the population estimate within 10 km of the estuary is 410,756 (Oak Ridge National Laboratory and UT-Battelle LLC 2008). Almost all of this population is in the greater Machala area and additionally includes people in extreme northern Peru.

Of the eight Ecuadorian estuaries analyzed, Grande Estuary is one of two where population pressure likely led to substantial mangrove forest loss. Mangrove forests exist along much of the mainland coastline of the province, throughout the estuaries that exist between the mainland and the numerous islands, and in the river mouths of the many rivers that enter the estuarine environment. As opposed to all other study areas, this province has a substantial ratio of fringe mangrove forests bordering the open ocean, although most of the mangrove forests are in sheltered estuaries. The dominant mangrove species as reported in the Jambeli Archipelago is again *Rhizophora mangle* at a 95% cover level with a small amount of *Avicennia germinans* also present at around 5% (Bravo and Altamirano 2006).

3.8.1 Climatic & Environmental Conditions

All the estuaries of Elo Oro Province reside in the Köppen-Geiger climate category Aw, known as Equatorial and winter-dry (Kottek et al. 2006). The average rainfall is 569 mm with a prolonged dry season from May to December; the average daily temperature is 26 °C (Bodero 1993). The minimum sea surface temperature is experienced in August with the water temperatures between 21.2 and 24 °C while the maximum sea surface temperatures are experienced in April with temperatures between 23.5 and 27.6 °C. The largest known tidal range at Puerto Bolivar is 3.51 m (NOAA Center for Operational Oceanographic Products and Services 2018).

3.8.2 Socioeconomic Conditions & Livelihoods

El Oro Province has a variety of community types residing on the estuary varying from tourist destinations, port cities, agricultural towns, and numerous small and isolated traditional fishing communities. The city of Machala, the provincial capital and the fifth largest city in the country, resides on an estuary with mangrove forest (Fig. 3.12). The city is known for agriculture in its hinterlands; it has been called the banana capital of the world. Although a large city, it only has 10% of the population of Guayaquil and does not dominate El Oro Province in the same manner as the city of Guayaquil dominates Guayas Province. Just short distances from Machala are traditional fishing communities with only mangrove forest livelihood options available to them. Without a road connecting them to the mainland, these communities are isolated on the many small islands along the coast.

In the extreme southern portion of the El Oro study area around Grande Estuary, as many as seven villages were historically sustained by the concha negra industry with nearly 500 cockle collectors residing on the islands, and at least 300 additional harvesters traveling there daily from Puerto Bolivar, Puerto Jeli, and Puerto Hualtaco on the mainland (MacKenzie 2001). Many of the harvesters from the mainland were migrants whose harvesting areas in the north were destroyed during the 1980s and 1990s by shrimp farm construction (Altamirano et al. 1998). Indeed, this story is told by cockle collectors in Cayapas-Mataje who migrated to the area from Muisne and Cojimies when their ancestral harvesting grounds were lost to aquaculture. Ninety-six percent of cockle gatherers among the mangrove forests of El Oro report the recent loss of cockling ground due to shrimp farm expansion or not obtaining a concession, and a little over half report conflict with a shrimp farmer over access to cockling grounds (Rosero Moya and Santillan Salas 2011).

The traditional fishing and cockle gathering lifestyle is not contained to the 1980s, 1990s, or 2000s in El Oro Province. Cockle harvesting and the traditional fishing lifestyle still dominate in large portions of the Elo Oro study site. For example, the cockling and fishing industries still dominate the small fishing community of Isla Costa Rica close to the Peruvian Border (Fig. 3.12). Seventy percent of the village's 320 residents work fulltime in the cockle collecting industry and an average of 254 cockle collectors per day work the mangrove forest grounds nearby in Puerto Hualtaco (Beitl 2011). This region of southern El Oro province reports levels of poverty often greater than 90% (Collins 2010), emphasizing the lack of other opportunities should cockling become unviable in this region.

3.8.3 Biodiversity

Limited biodiversity information exists for the mangrove forests of El Oro Province. It can be assumed that similar flora and fauna exist to that of the eastern portion of the Guayas study area as the mangrove forest forms a singular almost complete geography along this coastline (Figs. 3.10 and 3.12). The mollusk, crustacean, and wild fisheries are all active in the Jambeli Archipelago, and therefore it can be assumed that many important fisheries remain. For example, in and around the community of Costa Rica seven mollusk species have been identified including the commercially important concha negra (Bravo and Altamirano 2006).

3.8.4 *Protected Status*

Only a small portion of the mangrove forests and estuaries of El Oro Province are MAE protected areas. A former military base located on the Peruvian border became Reserva Ecológica Arenillas in 2001, and this reserve contains a small portion of mangrove forest at its extreme northern extent (Fig. 3.12). The reserve primarily protects tropical dry forest and not mangrove forests. The coastal portion of the Reserva Ecológica Arenillas is only 1539 ha or 12% of the 13,170 ha total reserve area. As with other MAE reserves in Guayas Estuary, Cojimies Estuary, and Muisne, the delineation of the reserve does appear to intentionally follow the mangrove forest border in this region to purposefully capture the mangrove forest to the west of the small and isolated community of Puerto Pitahaya. Reserva Ecológica Arenillas is challenging to access, does not offer the facilities of many other SNAP sites on the coast, and users must coordinate with the MAE before attempting to enter the park. Counting only the coastal portion of Reserva Ecológica Arenillas, the percentage of protected estuarine and mangrove forest area within the Elo Oro study area (Fig. 3.12) is only 1.5%, the next lowest level of federal protection of all the study sites after Isla Puná.

In addition to the Reserva Ecológica Arenillas, a tiny portion of the isolated sliver of the much larger Macizo del Cajas Biosphere Reserve (Jiménez Tapia et al. 2013) runs offshore of the El Oro Province study area (Fig. 3.12). This portion of the park appears to omit the mangrove forest intentionally and captures a non-forested marine environment along with a narrow strip just offshore (Fig. 3.12). Although lacking any meaningful federal conservation protection and any Ramsar recognition, other federal programs aimed at mangrove forest protection are active in the El Oro study area. The Socio Manglar program is active in El Oro Province, empowering local people with protection of the mangrove forest. As of 2017, El Oro Province has 21 *custodia* agreements in place covering 11,318 ha of mangrove forest or 75% of the total mangrove forest in the province (Rodríguez 2018). This program pays *custodia* to manage the mangrove forests sustainably. Such localized agreements likely help mitigate the lack of federal protection or Ramsar recognition of mangrove forests and estuaries along the length of El Oro Province.

3.9 The Legal Framework of Ecuadorian Mangrove Forests and Estuaries

As early as 1978, the Ecuadorian Forest Agency (Decree 2939-B) issued regulations prohibiting the clearing of mangrove forest for shrimp farms and established a regulated system of mangrove forest clearing (Robadue 1995). The Ecuadorian Navy was placed in charge of enforcing these rules, but the Navy had limited or no presence in the estuarine environments, particularly near the interior, and although some inhabitants did seek Navy approval for their mangrove forest activities (Robadue 1995), the decree was never enforced in any meaningful way. Technically, by 1985 all Ecuadorian mangrove forests had full legal protection with Decree 825 of June 1985 stating that all mangrove forests in the nation were to be conserved, protected, and restored (Robadue 1995; Bravo 2007), but unfortunately, once again, this well-meaning government policy was never enforced in the estuaries.

By the 1990s, the seven agencies with responsibility for the mangrove forests and coastal estuaries combined their resources and created ZEMs for portions of five of the study sites in Ecuador via a mechanism named the Coastal Resources Management Program (PMRC) which was formed by presidential decrees in 1989 and 1982 (Coello et al. 1993; Robadue 1995). These ZEMs appear to have had more influence than earlier decrees, most likely because they involved the residents and community groups in their implementation and planning, but again, through this period, the estuaries' health declined and the mangrove forest clearing continued. These early failings of mangrove forest conservation legislation in the 1970s and 1980s, and somewhat in the 1990s, in addition to the declining environmental condition of the estuaries, likely led to more meaningful protection that occurred post-2000, spurred by MAE with a direct mandate to protect the coastal environment and livelihoods of the traditional coastal residents.

During the 2000s, the MAE became the primary government actor attempting to conserve the mangrove forests and coastal estuaries of Ecuador. Without the power or the desire to issue decrees, the MAE undertook a more community-based approach to mangrove forest conservation and estuarine regulation than earlier epochs under differing agencies. In 2000, the MAE began a process of assigning decade-long community managed concessions to *custodia* of the mangrove forest under the 'ancestral rights' and 'traditional user groups' mandate, and by the end of the decade more than 42 different communities held concessions on 37,818 ha of mangrove forests (Beitl 2012; Rosero Moya and Santillan Salas 2011). However, the process of assigning *custodias* may have itself resulted in conflict and livelihood disputes between those with concessions and those without (Rosero Moya and Santillan Salas 2011). Mangrove forest *custodias* that agree to use mangrove forests sustainably can enter the Socio Manglar program, which is part of the wider successful and internationally recognized Socio Bosque program (de Koning et al. 2011) that makes payments to *custodias* to maintain forests. As of 2017, 23,000 ha of mangrove forest is in the Socio Manglar cash for conservation system (Rodríguez 2018).

In 2018, large portions of the estuarine study areas had some level of federal protection with some level of enforcement (Beitl 2011). The study areas contain part or all of nine federally protected areas (Ministerio del Ambiente 2018), part or all of four Ramsar sites (RAMSAR 2006), and one biosphere reserve (Man and the Biosphere Programme 2013). The combined protected area in the study site is 1018 ha, or only 1.5% of the entire study area. The nine MAE SNAP sites were created in 1979, 1985, 2001, two in 2002, 2003, 2007, 2008, and 2010, demonstrating the post-2000 impetus towards conservation of the mangrove forest resources.

Many of the protected areas are delineated from the boundary of the remaining mangrove forest; that is, the protected area boundary purposefully meanders to capture individual mangrove stands. This form of delineation occurs in the Manglares Churute Ecological Reserve, Refugio de Vida Silvestre Manglares el Morro, La Reserva de Producción Faunística Manglares el Salado, and to a lesser extent Área Nacional de Recreación Playas de Villamil. Other protected sites in Muisne and Guayaquil, under the guise of the Vida Silvestre Manglares of the Rio Muisne Estuary Refuge (Fig. 3.5), are merely the isolated mangrove stands that remain within a larger delineated area.

Traditionally all lands below high tide in Ecuador cannot be owned, and the Navy is responsible for activities that occur within the mangrove forests including management and regulation of the mangrove forests (Robadue 1995). This management regime was rarely enforced, and this lack of concern may have contributed to the mangrove deforestation and shrimp farm expansion that occurred as the Navy had little interest in such environmental change. The Ecuadorian Ministry of Agriculture, Livestock, Aquaculture, and Fisheries has traditionally viewed their role as managing the mangrove forest resources, but in recent times they have been superseded by the MAE as the mangrove forests and the estuaries in which they reside, have come under the protected area umbrellas and are incorporated into SNAP. Additionally, Ramsar and UN biosphere recognition have brought the management of some estuaries into the international realm.

In 2008, Ecuador became the first country to enshrine full rights for nature in their constitution (Greene 2011; Kauffman and Martin 2017). Chapter 7, Rights for Nature, of the Ecuadorian Constitution, now states[1]:

- Art. 71. "***Nature or Pachamama, where life is reproduced and exists, has the right to exist, persist, maintain and regenerate its vital cycles, structure, functions and its processes in evolution.*** *Every person, people, community or nationality, will be able to demand the recognition of rights for nature before the public organisms. The application and interpretation of these rights will follow the related principles established in the Constitution. The State will motivate natural and juridical persons as well as collectives to protect nature; it will promote respect towards all the elements that form an ecosystem.*"
- Art. 72. "***Nature has the right to restoration.*** *This integral restoration is independent of the obligation on natural and juridical persons or the State to indemnify the people and the collectives that depend on the natural systems. In the cases of severe or permanent environmental impact, including the ones caused by the exploitation of nonrenewable natural resources, the State will establish the most efficient mechanisms for the restoration and will adopt the adequate measures to eliminate or mitigate the harmful environmental consequences.*"
- Art. 73. "***The State will apply precaution and restriction measures in all the activities that can lead to the extinction of species, the destruction of the ecosystems or the permanent alteration of the natural cycles.*** *The introduction of organisms and organic and inorganic material that can alter in a definitive way the national genetic patrimony is prohibited.*"
- Art. 74. "***The persons, people, communities and nationalities will have the right to benefit from the environment and form natural wealth that will allow wellbeing.***

[1] Translation provide by https://therightsofnature.org/

The environmental services cannot be appropriated; its production, provision, use, and exploitation, will be regulated by the State."

Ecuadorian mangrove forests will likely benefit from the Rights of Nature now incorporated into the Ecuadorian constitution, although the mechanisms to implement the constitution on the ground are quite vague. Additionally, the MAE footprint over the estuaries continues to grow, and new environmental reserves are being created at regular intervals. As with the conservation of the past, success will likely be determined by community buy-in to these projects and meaningful enforcement of mangrove forest and wider conservation regulations. As opposed to earlier ministries' roles in conserving the mangrove forests which were often vague, in conflict with other sections of their wider mission, or seen as an outright inconvenience, the MAE are well positioned to succeed in their mandate of conserving the remaining mangrove forests and restoring the health of the coastal estuaries.

3.10 Study Area Summary

The study areas in this book about the Colombian border in the extreme north (Fig. 3.1, Study Area i) and abut the Peruvian border in the extreme south (Fig. 3.1, Study Area vii). The sites cover all the important estuarine environments along the coast and are contained in four of the five historic coastal provinces with mangrove forest. The combined study area size is 487,854 ha and contains greater than 99% of the 1969 known mangrove forest cover on mainland Ecuador.

The 2015 population in the combined study areas is 1,310,237, within 5 km of the study area it is 2,237,915, and within 10 km of the study area it is 3,009,822. The study sites either partially contain or are near the largest Ecuadorian city of Guayaquil, the fifth largest city of Machala, and the sixth largest city of Durán. Other major or smaller towns close to the study sites include San Lorenzo and Muisne in Esmeraldas Province, and Pedernales and Bahia de Caráquez in Manabí Province. Hundreds if not thousands of small communities dot the estuaries, a few of these towns and their livelihoods are documented above, but the majority are not. In these estuarine communities, traditional fisheries-based livelihoods used to dominate, and in many cases, still do today. Mangrove forest-related activities such as tourism and ecotourism have begun to supplement the livelihoods of these small communities.

Though often supplemented by tourism and agriculture, outside of the major cities traditional fishing remains the dominant livelihood option. The role of traditional fisheries has important gender dynamics, particularly in the Esmeraldas study area. Much of the concha negra and other cockle fisheries, particularly in the north, was reserved for women (Ocampo-Thomason 2006). Although the concha negra harvesting occurring in the mangrove forests should not be eulogized, concha negra values are generally high enough to provide a good living to harvesters (Ocampo-Thomason 2006; MacKenzie 2001). This fishery is in extreme peril and is all but extinct in four of the study sites (ii, iii, iv, v) and only thrive in one site (i). Concha negra only exists in any meaningful abundance in the roots of the *Rhizophora mangle*, so as mangrove forest clearing continues, the concheras and concheros are left without a sustainable livelihood.

The broader stories of the study areas have many positive components, particularly the responses by local communities to environmentally driven mangrove forest loss. The eco-city of Bahia de Caráquez has undergone a decade-long mangrove forest replanting initiative to restore the native fisheries. The artisanal residents of Cayapas-Mataje successfully resisted shrimp farm expansion and mangrove deforestation and established a national park with the express purpose of protecting their interests. Much of the Guayas Estuary has come under protected status as part of the SNAP Program with the MAE responsible for six protected areas in the Guayas Estuary. Five of these protected sites were created post-2000, again with the intent of protecting the mangrove forests and the livelihoods and food security of artisanal residents who rely on healthy mangrove forests and estuaries. In El Oro province the community has mobilized to conserve their mangrove forest resources; community actors have taken control of much of the fishery, numerous groups are actively replanting mangrove forest, and small portions of the mangrove forest are now protected.

The Ecuadorian government has taken the time to conduct the analyses and application processes to have four of the mangrove forest sites internationally recognized as Ramsar sites and have begun the process of establishing numerous marine reserves in the nearshore coastal waters. One part of the study area is now part of a UN-recognized biosphere reserve. Recognition as a SNAP reserve, biosphere reserve, Ramsar site, or marine protected area provides numerous additional benefits to mangrove forests beyond the actual delineation and enforcement of the protected area. For example, MAE protected status requires the creation and maintenance of detailed management plans and much of the biodiversity and livelihoods information contained within this manuscript comes from the regular management plans conducted as part of the MAE protected area process. These reports have intrinsic value espousing the biodiversity and livelihood conditions of residents in and around the SNAP site. Ramsar and biosphere reserve status both require a submission process that quantifies the importance of the local environment and contains commitments to on-site maintenance. Again, these important items, although only procedural to the creation of the protected status, produce meaningful information that is available to researchers, local community actors, NGOs, and the government, providing better knowledge of the study areas and the lifestyles of residents near them. Finally, the protected status brings Ecuadorian and international visitors to these sites, increasing awareness of the important estuarine environments and encouraging preservation of the resource to continue to attract visitors.

Recommended Further Reading

Cayapas-Mataje Estuary Site Information and Traditional Livelihoods

Ocampo-Thomason (2006)

Muisne and Cojimies Estuaries Site Information and Traditional Livelihoods

Capitulo (2008)

Chone Estuary Site Information and Traditional Livelihoods

Coello et al. (1993)
Valderrama (2014)

Guayas Estuary Estuary Site Information and Traditional Livelihoods

Maldonado et al. (2014)
Yanez (1996)
Ministerio del Ambiente del Ecuador (2010)

Isla Puná Estuaries Site Invormation and Traditional Livelihoods

Elizaldes Consultora (2015).

El Oro Province Site Invormation and Traditional Livelihoods

Beitl (2012)
Rodríguez (2018)

Mangrove Legislation

Robadue (1995)

Flora and Fauna of Isla Puná Isla Puná Site Information

Madsen et al. (2001)

MAE Protected Areas

Ministerio del Ambiente (2018)

Concessions and Socio Bosque

de Koning et al. (2011)

Mangroves Across all of Coastal Ecuador

Bodero (1993)

Data Availability

The study areas are available for download from https://doi.org/10.7910/DVN/7U8OHW.

The protected area data for each study site are available for download from https://doi.org/10.7910/DVN/ZVBFOM

A suite of 1:50,000 topographic maps (supplemented with 1:25,000 for specific areas) for each study site and the wider Ecuadorian coastline are available at https://doi.org/10.7910/DVN/21FMNQ

References

Altamirano M, Bravo M, Elao R, Noboa R, Rosero J (1998) Otorgacion, en custodia, de un area de manglar a un grupo de usuarios en el estero Saca Mano. Maestria Manejo de Recursos Costeros (Convenio PMRC ESPOL) Modulo III Economia de Recursos Instituto De Pesca, Guayaquil, Ecuador

Arriaga L, Montaño M, Vásconez J (1999) Integrated management perspectives of the Bahia de Caráquez zone and Chone River estuary, Ecuador. Ocean Coast Manag 42(2–4):229–241. https://doi.org/10.1016/s0964-5691(98)00055-6

Beitl CM (2011) Cockles in custody: the role of common property arrangements in the ecological sustainability of mangrove fisheries on the Ecuadorian coast

Beitl CM (2012) Shifting policies, access, and the tragedy of enclosures in Ecuadorian mangrove fisheries: towards a political ecology of the commons. J Polit Ecol 19:94–113

Blanchard J, Prado G (1995) Natural regeneration of Rhizophora Mangle in Strip Clearcuts in Northwest Ecuador. Biotropica 27(2):160–167

Bodero A (1993) Mangrove ecosystems of Ecuador. In: Lacerda LD (ed) Conservation and sustainable utilization of mangrove forests in Latin America and Africa regions. ISME, Okinawa, pp 55–74

Bravo M (2007) Analisis de los acuerdos de uso sustentable ortogados a usuarios tradicionales para la protección y manejo del ecosistema manglar: Estudios de caso de las Concheras 18 de Octubre (Provincia Esmeraldas); Cangrejeros Seis de Julio (Provincia Guayas) y Mariscadores Costa Rica (Provincia el Oro). Programa de manejo de recursos costeros, Procedimiento CCI No PMRC-035-2005 Actualización de los planes de manejo de las concesiones del manglar

Bravo M, Altamirano M (2006) Actualización del Plan de Manejo del Manglar Concesionado a la Asociación De Pescadores Mariscadores y Afines De Costa Rica, Archipiélago de Jambelí. Programa de Manejo de Recursos Costeros, vol 1. PRMC, Guayaquil

Brinkhoff T (2011) City population statistics. URL: http://www.citypopulation.de. Last accessed 5 Sept 2012

Capitulo I (2008) Plan de Manejo Participativo y Comunitario del Refugio de Vida Silvestre del Estuario de Manglares Muisne-Cojímies: Diagnóstico Socioeconómico-Cultural. Ministry of the Environment, Quito

CLIRSEN (1987) Estudio Multitemporal de los Manglares, Camaroneras y Areas Salinas de la Costa Ecuatoriana, vol 1. Centro De Levantamientos Integrados De Recursos Naturales Por Sensores Remotos, Quito

CLIRSEN (2007) Actualizacion Del Estudio Multitemporal de Manglares, Camaroneras Y Areas Salinas En Las Costa Continental Ecuatoriana Al Ano 2006, vol 1. Centro De Levantamientos Integrados De Recursos Naturales Por Sensores Remotos, Quito

Coello S, Proafio-Lerowr D, Robadue D (1993) Special area management planning in Ecuador's Rio Chone estuary. Paper presented at the coastal zone, New Orleans, LA

Collins S (2010) Mangrove destruction and shrimp aquaculture in Ecuador: a focus on property right enforcement. University of Ottawa, Ottawa

Crawford B (2010) The Cojimies estuary in Ecuador. Coastal Resources Center: Graduate School of Oceanography. University of Rhode Island. http://www.crc.uri.edu/download/Ecuador_Cojimies_Program_Brief.pdf. Accessed 12 Jan 2010

Cucalón E (1983) Temperatura, salinity and water mass distribution off Ecuador during an El Niño event in 1976. Revista de Ciencias del Mr y Limnología INP Ecuador 2(1)

Cucalón E (1989) Oceanographic characteristics off the coast of Ecuador. A sustainable shrimp mariculture industry for Ecuador. University of Rhode Island, Coastal Resources Center, Narragansett

de Koning F, Aguinaga M, Bravo M, Chiu M, Lascano M, Lozada T, Suarez L (2011) Bridging the gap between forest conservation and poverty alleviation: the Ecuadorian socio Bosque program. Environ Sci Pol 14(5):531–542. https://doi.org/10.1016/j.envsci.2011.04.007

Elizaldes Consultora (2015) Actualización del Plan de Desarrollo y Ordenamiento Territorial Puná 201–2019. Quito Ecuador
Fernandez J, Yepes H (1998) Learning from Earthquakes: The Bahia de Caraquez Earthquake of August 4th, 1998. EERI Special Earthquake Report 1998. Earthquake Engineering Research Institute, Oakland
Fick SE, Hijmans RJ (2017) WorldClim 2: new 1-km spatial resolution climate surfaces for global land areas. Int J Climatol 37(12):4302–4315
Greene N (2011) The first successful case of the rights of nature implementation in Ecuador. The Global Alliance for the Rights of Nature Available: http://www.therightsofnatureorg/first-ron-case-ecuador
Hamilton SE (2012) The Impact of Shrimp Farming on Mangrove Ecosystems and Local Livelihoods along the Pacific Coast of Ecuador, vol 1. Dept. of Geography and Geology, Doctoral Dissertation, vol 3477166, 1st edn. ProQuest, UMI Dissertation Publishing, Hattiesburg, Mississippi
Hamilton SE, Fitt J, Kaempf C, Coyne SJ (2015) The impacts of sedimentation in the Río Chone estuary. In: Toulkeridis T, Andrade L, Mejía M, Cartay R, Regalado L (eds) VIII Jornadas Académicas Turismo y Patrimonio, Compartiendo lo nuestro con el mundo. Memorias Contribuciones Científicas., vol 1. Descubriendo el Geoturismo en Manabí: Distrito Jama – Pedernales, vol 1. ESPAM MFL, Calceta, pp 38–43
Hamilton SE, Lovette JP, Borbor-Cordova MJ, Millones M (2017) The carbon holdings of northern Ecuador's mangrove forests. Ann Am Assoc Geogr 107(1):54–71. https://doi.org/10.1080/24694452.2016.1226160
Healy T, Wang Y, Healy J-A (2002) Muddy coasts of the world: processes, deposits and function, vol 4. Elsevier, Amsterdam
Herrera M, Elao R (2007) Analisis de Amenazas a La Biodiversidad en el Estuario De Cojimies (Ecuador): Identificacion de Amenazas y Oportunidades Para El Manejo del Estuario de Cojimies y Su Area de Influencia Dentro del Programa SUCCESS. Sustainable Coastal Communities and Ecosystems (SUCCESS), a Component of the Global Integrated Management of Coastal and Freshwater Systems (IMCAFS) Program. USAID, EPP-A-00-04-00014-00
IUCN SSC (2010) IUCN red list categories and criteria: version 3.1
IUCN, UNEP (2016) The World Database on Protected Areas (WDPA). www.protectedplanet.net. Accessed 1 Dec 2018
Jiménez Tapia AK, Rueda R, Quelmita R (2013) Incidencias de la declaración del macizo del Cajas como reserva del biósfera en el aspecto turístico del Parque Nacional Cajas
Kauffman CM, Martin PL (2017) Can rights of nature make development more sustainable? Why some Ecuadorian lawsuits succeed and others fail. World Dev 92:130–142
Kottek M, Grieser J, Beck C, Rudolf B, Rubel F (2006) World map of the Köppen-Geiger climate classification updated. Meteorol Z 15(3):259–263
Leon FD, Johnson S, Marcon J, Phares N (2016) Watershed management to address sedimentation of Ecuador's Rio Chone estuary. University of California, Santa Barbara
MacKenzie CL (2001) The fisheries for mangrove Mockles, *Anadara* spp., from Mexico to Peru, with descriptions of their habitat and biology, the Fishermen's lives, and the effects of shrimp farming. Mar Fish Rev 63(1):39
MacKenzie CL Jr (2001) The fisheries for mangrove cockles, Anadaraspp., from Mexico to Peru, with descriptions of their habitats and biology, the Fishermen's lives, and the effects of shrimp farming. Mar Fish Rev 63(1):1–39
Madsen JE, Mix RL, Balslev H (2001) Flora of Puná island: plant resources on a neotropical island. Åarhus University Press, Aarhus
Maldonado M, Torres J, Carvajal R (2014) Plan de Fortalecimiento para el Desarrollo Sustentable del Sitio Ramsar Manglares del Estuario Interior del Golfo de Guayaquil "Don Goyo". Ministry of the Environment, Quito

Man and the Biosphere Programme (2013) Macizo del Cajas. UNESCO. http://www.unesco.org/new/en/natural-sciences/environment/ecological-sciences/biosphere-reserves/latin-america-and-the-caribbean/ecuador/macizo-del-cajas/. Accessed 12 Jan 2018

Minget P (1741) Guayaquil; . Compendio Histórico de la Provincia, Partidos, Ciudades, Astilleros, Ríos, y Puerto de Guayaquil en las Costas de la Mar del Sur – Año de 1741 – Dedicado al Rey Don Felipe Quinto Nuestro Señor en su Real y Supremo Consejo de las Indias. Spain

Ministerio del Ambiente (2018) National System of Protected Areas. Ministerio del Ambiente. http://areasprotegidas.ambiente.gob.ec/es/todas-areas-protegidas-por-region?t=C. Accessed 12 Jan 2018

Ministerio del Ambiente del Ecuador (2010) Plan de Manejo del Refugio de Vida Silvestre Manglares El Morro. Fundación Ecuatoriana para el Estudio de Mamíferos Marinos (FEMM), Fundación …

NOAA Center for Operational Oceanographic Products and Services (2018) Tides & currents. NOAA. https://tidesandcurrents.noaa.gov/. Accessed 11 Jan 2018

Oak Ridge National Laboratory, UT-Battelle LLC (2014) LandScan 2000–2012 (trans: Energy USD, Defense USDo). East View Geospatial, Oak Ridge

Ocampo-Thomason P (2006) Mangroves, people and cockles: impacts of the shrimp-farming industry on mangrove communities in Esmeraldas province, Ecuador. In: Hoanh CT, Tuong TP, Gowing JW, Hardy B (eds) Environment and livelihoods in tropical coastal zones: managing agriculture-fishery-aquaculture conflicts (comprehensive assessment of water Management in Agriculture Series), vol 2. Oxford University Press, London, pp 140–153

Ochoa E (1997) Majagual: the tallest mangroves in the world. Intercoast Network (1):17

Odum H, Odum B, Campbell D, Olsen S (1991) Un Reconocimiento de Temas Ambientales en la Región Costera y Zonas Especiales de Manejo Seleccionadas. Programa de Manejo de Recursos Costeros (PMRC) Convenio de Cooperación Técnica entre el Gobierno de Ecuador, Universidad de Rhode Island y USAID 91p (Serie Estudios No 3)

Ramsar Sites Information Service (2004) Wetlands. Accessed 2 Jan 2008

Robadue DD (1995) Eight years in Ecuador: the road to integrated coastal management, vol 2088. Coastal Resources Center, University of Rhode Island Narragansett

Rodríguez FVL (2018) Mangrove concessions: an innovative strategy for community mangrove conservation in Ecuador. In: Makowski C, Finkl CW (eds) Threats to mangrove forests: hazards, vulnerability, and management, vol 25. Springer, Cham, pp 557–578

Rosero Moya K, Santillan Salas C (2011) Ministerio del ambiente entrega acuerdo de uso sustentable y custodia de manglar. Boletin No 498 09-06-2011

Souris M (2008) Ecuador: Digital Elevation Data. Ecuador, vol 2008. School of Engineering and Technology, Asian Institute of Technology, Pathumthani

Spalding M, Kainuma M, Collins L (2010) World atlas of mangroves. Earthscan, London

Stevenson MR (1981) Seasonal variations in the Gulf of Guayaquil, a tropical estuary. Boletin Cientifico y Téchnico INP 4:1–133

Stram DL, Kincaid CR, Campbell DE (2005) Water quality modeling in the Rio Chone estuary. J Coast Res 21(4):797–810. https://doi.org/10.2112/011-NIS.1

Strategic Framework and Guidelines for the Future Development of the List of Wetlands of International Importance of the Convention on Wetlands (2006) Ramsar, Iran. http://www.ramsar.org/cda/en/ramsar-documents-guidelines-strategic-framework-and/main/ramsar/1-31-105 percent5E20823_4000_0__. Accessed 2 Jan 2008

Twilley RR, Cárdenas W, Rivera-Monroy VH, Espinoza J, Suescum R, Armijos MM, Solórzano L (2001) The Gulf of Guayaquil and the Guayas River estuary, Ecuador. In: Seeliger U, Kjerfve B (eds) Coastal marine ecosystems of Latin America. Springer, Berlin/Heidelberg, pp 245–263. https://doi.org/10.1007/978-3-662-04482-7_18

Ubilla IT (1998) Variaciones morfológicas y batimétricas de la línea de costa en el estuario del río Chone, producidas por los eventos ENSO. Bulletin de l'Institut français d'études andines 27(3):557–563

Valderrama NSH (2014) Plan DeEManejo Refugio de Vida Silvestre Isla Corazón y Fragata. Quito, Ecuador

Veach K (1996) Gender, resource use, conservation attitudes and local participation in mangrove fishing villages in northern Esmeraldas province, Ecuador. University of Florida, Gainesville

Yanez B (1996) Mangrove ecological reserve Churtue management plan – phase 1, vol IV. Ministerio del Ambiente, Quito

Chapter 4
Assessing 50 Years of Mangrove Forest Loss Along the Pacific Coast of Ecuador: A Remote Sensing Synthesis

Abstract To assess the changes in landuse and landcover resulting from the shrimp farm expansion occurring in each of the study areas, a remote sensing analysis of each estuary in Ecuador is conducted. This remote sensing analysis documents mangrove forest landuse and landcover (LULC) change as well as shrimp farm LULC change that occurred in each of the estuaries. Also, the remote sensing analysis pinpoints the approximate time the LULC transition occurred and the exact location of these LULC changes. The remotely-sensed data has the required spatial, temporal, and radiometric resolutions to detect if shrimp farms directly displaced mangrove forest at the sub-estuary level. This remote sensing analysis can then be used in Chap. 5 to answer the critical question of how much aquaculture-driven LULC change has occurred at the local level since commercial shrimp farms arrived in each estuary in Ecuador. The primary earth observing system (EOS) used to conduct the LULC change analysis is the Landsat program as it has the required resolutions to achieve these stated goals. This chapter describes a repeatable and widely-applicable method based on remotely sensed data synthesis that can be applied to other locations or even other landocovers. All the input, processing, and output data used or generated are provided in an accompanying website.

Keywords EOS · Landsat · Aster · Rapid eye · Remote sensing · Topographic maps · Air photos · Map projections · Land use · Land cover · Land use change · Land cover change

4.1 Remotely Sensed Survey Data Scene Selection

For each study area (Fig. 3.1), the Landsat archives were visually examined to determine the first appearance of commercial shrimp farming in each study area. The early large and extensive shrimp growout ponds appear as highly regular

Electronic Supplementary Material The online version of this chapter (https://doi.org/10.1007/978-3-030-22240-6_4) contains supplementary material, which is available to authorized users.

S. E. Hamilton, *Mangroves and Aquaculture*, Coastal Research Library 33,
https://doi.org/10.1007/978-3-030-22240-6_4

rectangular patterns in the estuaries and are quite easy to recognize from space as no other estuary LULC has this shape or size (Fig. 2.3). Once the appearance of the first commercial shrimp farms is ascertained from the Landsat archives, then the earliest Landsat image with no commercial shrimp farming and with suitable atmospheric conditions is selected as the base-level dataset for that estuary. This image would typically be the Landsat scene immediately before the appearance of the first commercial shrimp ponds with suitable atmospheric conditions. Therefore, the initial remotely-sensed survey period is not a singular date or even a distinct year, but it is an event, and that event is the first appearance of commercial ponds in the estuary. Hence, the initial survey date can differ from estuary to estuary but not within an estuary.

For the three study sites of Isla Puná, Isla Puná North, and Cayapas-Mataje respectively (Fig. 3.1), the Landsat images are from 1973, 1985, and 1986 respectively and are used for the initial LULC surveys. In Chone Estuary, Guayas Estuary, El Oro Province, Muisne Estuary, and Cojimies Estuary (Fig. 3.1), some early commercial shrimp farms are visible in the earliest Landsat images available. For these areas, supplemental pre-Landsat data are required.

To meet the need for high-resolution pre-Landsat data for these five estuaries, topographic maps, and aerial imagery were obtained from the Instituto Geografico Militar (IGM) in Quito and the IGM in Guayas. The high-resolution aerial photography and topographic maps are used in the initial survey of each estuary. The topographic maps for Chone represent 1968 LULC (Table 4.1), the topographic maps for Cojimies and Muisne estuaries represent 1971 LULC (Table 4.1), and the topo-

Table 4.1 Survey years and instrument

	Base	1	2	3	4	5	6	7	8	Final	Total
Cayapas-Mataje	1986 TM	1997 TM	2006 TM	2014 OLI							4
Chone	1968 PhTo	1977 Ph	1984 TM	1991 TM	1998 TM	2001 ETM+	2004 ETM+	2008 AST	2011 REye	2014 REye	10
Cojimies	1970 to	1986 TM	1998 TM	2005 AST	2014 OLI						5
El Oro	1963 to	1985 TM	1997 TM	2006 AST	2011 AST	2014 OLI					6
Muisne	1970 to	1986 TM	1998 TM	2005 ETM+	2014 OLI						5
Isla Puná	1973 MSSTo	1985 TM	1990 TM	1997 TM	2008 AST	2014 AST					5
Isla Puná North	1977 To	1985 TM	1997 TM	2006 AST	2014 OLI						5
Guayas	1985 TM	1990 TM	1999 ETM+	2014 OLI							4

Caption – TM is the Thematic Mapper at 30 m resolution, AST is ASTER at 15 m resolution, OLI is the Operational Line Imager at 30 m resolution, MSS is the Multispectral Scanner 1 at 80 m resolution, Ph is 1 m Aerial Photography, To are 1:25,000 Topographic Maps, and REye is RapidEye at 5 m resolution. The total is the number of surveys per estuary graphic maps for El Oro represent 1963 and 1968 LULC with small areas supplemented with maps from 1971 and 1977 (Table 4.1). In Guayas estuary, the earliest cloud-free images from 1985 had shrimp farms present in the estuary, but no earlier images with suitable atmospheric and radiometric conditions could be found, and no suitable topographic maps could be obtained. For this reason, in Guayas, the 1985 landsat imagery with shrimp farms present is the base-level dataset (Table 4.1).

Approximately decadal longitudinal estuarine-level LULC surveys were conducted to document land use change from pre-aquaculture to the present day (Table 4.1). Due to data omissions and issues of cloud cover, each estuary was not sampled at exactly decadal intervals (Table 4.1). Chone Estuary was over-sampled across time (Table 4.1) and space to give insight into trends between the decadal periods with an average of two surveys per decade conducted (Table 4.1). Most surveys following the 1960s and 1970s air photo and topographic map surveys were Landsat-based (Table 4.1). The four Landsat sensors utilized are Landsat Multispectral Scanner (MSS) for a small area of Isla Puná for one survey, Landsat Thematic Mapper (TM) on Landsat 4 and Landsat 5, Landsat Enhanced Thematic Mapper Plus (ETM+) on Landsat 7, and Landsat Operational Line Imager (OLI) on Landsat 8 (Table 4.1). These instruments are described in-depth in Sect. 4.3.

During 2003, Landsat's ETM+ started experiencing scanning issues. All but one of the estuaries no longer had ETM+ data for analysis due to this error. During this period, Advanced Spaceborne Thermal Emission and Reflection Radiometer (ASTER) data mostly replaced Landsat TM in the study areas besides Muisne Estuary and portions of Chone Estuary where ETM+ could still be used. (Table 4.1). The Nature Conservancy donated ASTER imagery for the entire coastline of Ecuador; these data are now available free of charge through various (National Aeronautics and Space Administration) NASA portals. The estuary of Muisne and most of Chone Estuary fall inside the ETM+ small envelope of correct data and hence is utilized in this estuary post-2003 (Table 4.1). ASTER was the primary instrument from 2003 until the availability of Landsat 8 OLI data after 2012 (Table 4.1). Then again after 2013, Landsat, OLI, once again becomes the dominant sensor except for Chone Estuary where additional funding was obtained to purchase very high-resolution RapidEye 5 m resolution data (Table 4.1). These 5 m data allow for more granular insights of LULC changes occurring in Chone Estuary.

4.2 Integration of Historic Topographic Maps and Air-Photos with the Modern Satellite Data

Coastal Ecuador has an almost complete repository of topographic maps from the late 1960s and early 1970s, constructed at 1:25,000 scale. The topographic maps were initially created via visual photographic interpretation and field-based ground truthing of high-resolution aerial photography flown by the United States Air Force. As opposed to topographic maps of the same period in English-speaking countries in South and Central America (such as Belize and Guyana), countries with a Spanish colonial legacy (such as Ecuador and many others in the region), have mangrove forest delineated as a complete and distinct feature in their topographic maps. The mangrove forests are identified on the topographic maps by symbology that is distinct from terrestrial forests, mangroves forests are included in the legend, and the maps are delineated into individual mangrove forest areas.

The primary local data sources utilized for LULC delineation pre-Landsat are the 1:25,000 locally produced topographic maps. The topographic maps were purchased from the IGM in Quito by the author in 2007 and 2008. The topographic maps were scanned at 300 DPI by a large format map scanner that preserves scale and then the author converted the maps into high-resolution uncompressed digital images. The analog topographic maps are georeferenced initially using the Provisional South American Datum (PSAD) of 1956 within Universal Transverse Mercator (UTM) zone 17 north or zone 17 south, depending on their location relative to the equator. Therefore, the digital versions of the topographic maps contain a graticule in UTM 17 PSAD 56 but are not digitally located in any geographic space at this point, as the digital maps are merely scanned images containing no world file or other suitable automatic digital referencing information.

To allow for the accurate georeferencing of the scanned digital topographic maps and to convert them into a spatially aligned geographic information system (GIS) product, a 1000 m fishnet in UTM 17 PSAD 1956 was programmatically created. The fishnet spans the entire coastal region of Ecuador, as well as adjacent areas of the Pacific Ocean. The fishnet contains all of the 192,240 potential 1:25,000 topographic graticule x-axis and y-axis intersections in the southern hemispheric portion of Ecuador and the equivalent 42,720 intersections in the north. That is, the fishnet contains every possible location where the x-axis and y-axis graticule lines on any individual topographic map intersect.

The fishnet intersections were then converted into georeferenced points in a GIS. This fishnet structure creates 14 precise and exact reference points on the x-axis, and nine precise and exact reference points on the y-axis for each topographic map (Fig. 4.1). The raster topographic maps were then georeferenced in GIS, using the rubber-sheeting procedure, by matching the graticule intersections of the topographic map to each of the corresponding fishnet points (Fig. 4.1). The rubber-sheeting process is relatively straightforward as the scale of the topographic maps is preserved via the process outlined above, and no distortion is required. A root-mean-square error (RMSE) of under 1 m was achieved for all input maps, and no polynomial transformation was required; this process is often referred to as fitting in GIS. The entire topographic map reference process above was then repeated iteratively for each purchased topographic map until the entire estuary had topographic map coverage. The process was repeated for each estuary, aside from Guayas, Cayapas-Mataje, and Isla Puná for which Landsat suffices for the initial LULC delineation.

Although the topographic maps provide accurate and precise 1:25,000 LULC information, much of the map creation and feature delineation is up to the qualitative interpretation of the initial cartographers. To test the size of mangrove forest

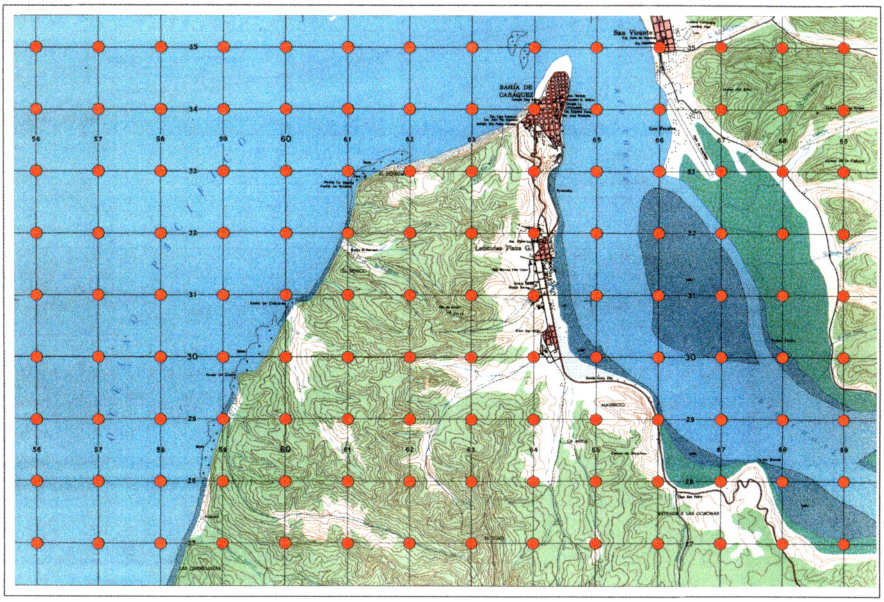

Fig. 4.1 Fishnet construction for a topographic map in Chone estuary
Caption – A 1:25,000 scanned topographic map from 1968 covering the coastal portion of the Chone Estuary in Ecuador after fitting. The red dots represent the fishnet points; each point has a coordinate pair in UTM 17 PSAD 56 that matches the graticule. The topographic map graticule line intersections were matched to each of their corresponding fishnet points. The matching ensures highly accurate georeferencing of the digital topographic map. Mangrove forest can be seen using the standard topographic mangrove forest symbol on both the north side of the estuary (68,31) and in more limited amounts of the south side (67,28). Note the small creeks delineated in the northern mangrove stand

included in the topographic maps, to ascertain the accuracy of mangrove forest delineation in the topographic maps, and to verify the actual resolution of the input data; source aerial photography is required. For the Chone Estuary, the author purchased both the1968 and 197 k, black and white aerial photography from which the topographic maps were created to test the topographic map creation and obtain a further longitudinal survey data point (Table 4.1). The original photographs were made available on traditional black and white silver halide crystal plates, with each plate measuring 23 cm on both the x-axis and y-axis. Complete metadata for these photographs is not available, although the scale and focal length are stamped on each photo. Each aerial photograph was scanned at 1400 μm and its scale calculated to represent a ground pixel resolution of 0.84 m (Fig. 4.2). This inclusion of aerial photography tests the accuracy and representativeness of the topographic maps against this higher resolution source data.

The aerial photography was resampled to 1 m resolution and georeferenced via rubber-sheeting using a first-order polynomial. The rubber-sheeting was achieved by finding identifiable features that exist both within the photography and on the

$$Pm = ((S \times \mu m))/1{,}000{,}000$$
$$Pm = ((60{,}000 \times 14))/1{,}000{,}000$$
$$Pm = .84$$

Fig. 4.2 Establishing a pixel size for aerial photography. Pm = pixel size in meters, S = Scale (1:60,000 would is represented as 60,000), µm = Scanning precision in micrometers (Jensen 2005)

Fig. 4.3 1968 aerial photography overlaid on topographic maps created from the aerial photography in the Chone estuary
Caption – The 1 m air photos overlaid on the topographic maps. The photos have a 10% transparency applied to view the matching of the mangrove stands depicted in the photographs and the topographic mangrove stands. The 1 m photographic mangrove forest delineations align almost entirely with the 1:25,000 aerial photography

ground (Fig. 4.3). The topographic mangrove forest delineation was found to be an almost exact match to the 1 m photography (Fig. 4.3) on which it was based. This testing of the accuracy of the 1:25,000 topographic maps using their source imagery confirmed that the LULC delineated in the topographic maps was comparable to the satellite remote sensing instruments that are used in later years. The 1977 aerial photographs were used to create an entirely new LULC delineation for this period in Chone Estuary. Due to cost and time limitations, the topographic maps were only tested in one estuary using supplemental aerial photography. It is assumed the remaining topographic maps for the other estuaries, which are from the same map series, the same flight, compiled using the same cartographers, and compiled via the same methods will have similar accuracy and representation characteristics.

As noted previously, the early topographic maps of Ecuador are georeferenced to a local datum named PSAD 1956. All modern satellite data is referenced to the

Earth-centered datum World Geodetic System (WGS) 1984. Spatial data in PSAD 1956 is not comparable to spatial data in WGS 1984; thus, any derived datasets from input data referenced in PSAD 1956 will not align with modern satellite data and could have inconsistencies of up to 300 m in Ecuador (NIMA 2000). To make matters more complicated, the P in PSAD means provisional as the datum was never entirely constructed, verified, or corrected across South America or Ecuador. Therefore, PSAD 1956 has many inconsistencies and errors that have never been addressed. To allow for the pre-satellite air photo and topographic map data to be compared against modern era data, it needs to be converted to WGS 1984. Due to the inconsistencies in PSAD 1956, no single geographic transformation works. Each estuary requires a unique geographic transformation to be constructed to allow for the transformation from PSAD 1956 to WGS 1984.

The first step in the transformation from PSAD 1956 to WGS 1984 is to apply an Ecuadorian-wide datum conversion to bring the two disparate datasets into much closer alignment. Two adjustments are provided for Ecuador by the National Geospatial Agency. These are known as PRP-E 1991 and PRP-M 1987 respectively (NIMA 2000). PRP-E was selected due to it being a more recent transformation, having substantially lower levels of uncertainty, and being Ecuador-specific as opposed to attempting to fit many nations across South America. Additionally, both transformations were tested on a sample of the topographic maps and air photos; and PRP-E brought the data into much closer alignment than PRP-M. PRP-E is a geocentric adjustment that applies a delta shift along the x-axis of (ΔX) 278 m and along the y-axis of (ΔY) 171 m, with calculated errors of $\pm$ 3 m and $\pm$ 5 m respectively (NIMA 2000). This adjustment was carried out in ESRI ArcGIS using a cubic convolution function for the raster images, converting the pre-satellite era to be almost comparable to WGS 1984 referenced data. This geographic adjustment is now prebuilt in ESRI software and can be retrieved by searching for PSAD_1956_To_WGS_1984_6.

The Ecuador-wide adjustment process got each of the estuary's topographic maps and air-photo data within a few meters of WGS 1984 alignment for each estuary and converted into WGS 1984. Such accuracy would likely suffice for 90 m, 60 m, and 30 m resolution Landsat-derived data comparisons, but further locational accuracy is needed to allow for comparison to modern VHR systems, such as RapidEye and Sentinel as well as to allow for comparison to high-resolution field surveys. Therefore, a more refined adjustment was required at the estuary level to bring these data into closer alignment.

The final geographic adjustment for each estuary was achieved by locating humanmade objects present in the early 1 m resolution imagery and the derived topographic maps that are still present and visible in the environment today. Such objects are most commonly road intersections or concrete docks. The features are then precisely located using Global Positioning System (GPS) in the field. The current WGS 1984 coordinate pair and the historic PRP-E adjusted WGS 1984 coordinate pair were then compared, and the differences in location between each point pair was calculated. The difference was generally under 3 m. By repeating this process across numerous well-distributed points in each estuary, a final minor shift could be applied at the estuary level after the PRP-E shift to complete the alignment process. The new estuarine level adjustment grids were created via interpolation from all input point pairs for each estuary, and the estuarine level shift was applied to align all historical data at the estuarine level.

The process of obtaining, scanning, georeferencing, and creating a ground control network for each of the topographic maps across all estuaries where they were used as the base image took a substantial amount of time. The process of verifying these topographic maps for Chone via aerial photography for the entire estuary took an additional extended period. Despite the time and cost constraints, this process allows the remote sensing analysis presented to continue back in time before the advent of satellite remote sensing while maintaining comparable spatial resolutions and feature extraction accuracy. The methods presented result in a remotely-sensed LULC analysis going back 50 years for the entire coast of Ecuador. This pre-satellite era data makes this mangrove forest analysis one of the most extended remotely-sensed time series ever conducted for mangrove forests at the national scale.

4.3 Remote Sensing Image Selection and Sensor Information

With little fanfare, the near-complete international Landsat archive was gradually made available for free via internet download by the United States Geological Survey (USGS) starting in 2007 (Woodcock et al. 2008). This release allowed for longer-term affordable remote-sensing analyses of the type conducted in this manuscript. The Landsat archive allows for near-continuous remote sensing analyses to be conducted back to late 1973. The Landsat program has been operational since late 1972 and provides a 45-year remote sensing history of Ecuador and most terrestrial locations globally (Fig. 4.4). In addition to Landsat, supplemental ASTER data are used in this analysis as well as commercial high-resolution RapidEye imagery. The primary remote sensing instruments are described below with reference to

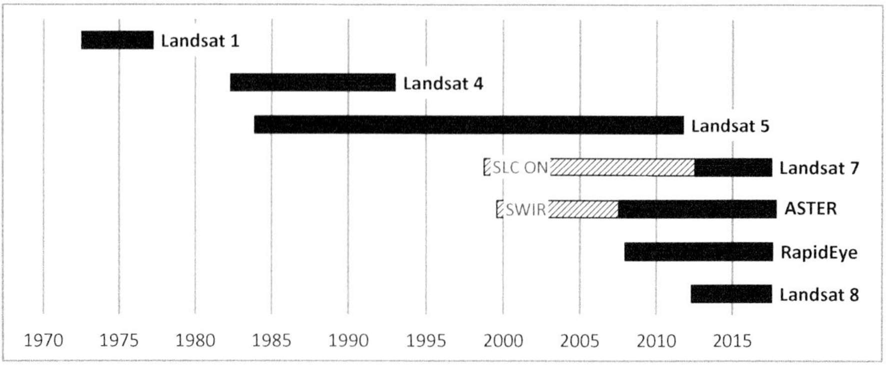

Fig. 4.4 Landsat, Aster, and RapidEye history

the continuity of both the radiometric and spatial resolutions used across all instruments and all estuaries named in Table 4.1.

4.3.1 MSS

Landsat 1 is used in this study only for LULC delineation in one portion of Isla Puná for one survey period. Landsat 1 is the first operational sensor in the Landsat program. Landsat 1 was launched in late 1972 with two operational sensors: The Return Beam Vidicon (RBV) and the MSS (Zanter 2016; Williams 2009; USGS and US Dept. of Interior 2018). The RBV sensor was decommissioned within a month of launch, and the MSS continued operation until 1978. The MSS location repeat cycle was 8 days with a scene size of 170 km by 185 km. The ground sampling interval or pixel size is 57 m by 79 m at the equator but is generally made available at a 90 by 90 m pixel size. The MSS collected data in four regions of the electromagnetic spectrum. These four regions are known as Band 4 which is in the visible green range from 0.5 μm to 0.6 μm, Band 5 which is in the visible red range 0.6 μm to 0.7 μm, Band 6 which is in the near-infrared range from 0.7 μm to 0.8 μm, and Band 7 which is in the near-infrared from 0.8 μm to 1.1 μm (Zanter 2016; Williams 2009; USGS and US Dept. of Interior 2018). The MSS path and row indicating the location of the imagery in the World Reference System (WRS)-1 or -2 for each estuary are included in Table 4.2.

Table 4.2 MSS, TM, ETM+, and OLI path and row information

	MSS path/row	TM ETM+ path/row
	WRS-1 day	WRS-2 day
Cayapas-Mataje	11/59	11/59
	10/59	10/59
Chone	11/60	11/60
	11/61	11/61
Cojimies	11/60	11/60
El Oro	11/62	11/62
	10/62	10/62
	10/63	
Muisne	11/60	11/60
Isla Puná W	11/62	11/62
Isla Puná N	11/62	11/62
Guayas	11/62	11/62
	11/61	11/61
	10/62	10/62

4.3.2 TM

Landsat 4 TM was in operation from mid-1982 to December 1993, and Landsat 5 TM was in service from March 1984 to November 2011 (Zanter 2016; Williams 2009; USGS and US Dept. of Interior 2018). Landsat 5 is used for most of the surveys in this manuscript (Table 4.1) and at decommissioning, was the longest running operational Landsat system. Both the Landsat 4 and the Landsat 5 TMs have the same characteristics and were essentially two of the same instruments. The ground sampling interval or pixel size for the relevant TM data is 30 m by 30 m at the equator. The TM repeat cycle was 16-days with a swath width of 185 km. As an enhancement on earlier Landsat systems, the TM collected data in six relevant regions of the electromagnetic spectrum (Table 4.3). Of these six ranges, three are in the visible spectrum, and the other three are outside the visible spectrum in the infrared region. The three regions in the visible spectrum are known as Band 1 which is in the blue range from 0.45 μm to 0.52 μm, Band 2 which is in the green range 0.52 μm to 0.60 μm, and Band 3 which is in the red range from 0.63 μm to 0.69 μm. The three infrared regions are known as Band 4 which is in the near-infrared range from 0.767 μm to 0.90 μm, Band 5 which is in the mid-infrared range from 1.55 μm to 1.75 μm, and Band 7 which is also in the mid-infrared range from 2.08 μm to 2.35 μm (Zanter 2016; Williams 2009; USGS and US Dept. of Interior 2018). Landsat TM follows the WRS-2 reference system. The TM path and row for each estuary are included in Table 4.2.

4.3.3 ETM+

As the name suggests, ETM+ is an enhancement of the earlier TM systems on Landsat 4 and Landsat 5 (Zanter 2016; Williams 2009; USGS and US Dept. of Interior 2018). The original ETM was part of the Landsat 6 mission but failed to

Table 4.3 Aster band information and landsat equivalents (Abrams et al. 2002)

System	Band	Name	μm	TM	ETM+	OLI
VNIR	1	Green	0.52–0.62	2	2	3
	2	Red	0.63–0.69	3	3	4
	3 V	Near infrared	0.78–0.86	4	4	5
SWIR	4		1.6–1.7	5	5	6
	5		2.145–2.185	7	7	6
	6		2.185–2.225	7	7	6
	7		2.235–2.285	7	7	6
	8		2.295–2.365	7	7	NA
	9		2.360–2.430	NA	NA	NA

reach orbit after launch in 1993. The replacement ETM onboard Landsat 7, known as ETM+, is essentially an update to the previous ETM with enhanced thermal imagery. The ETM+ was launched in April 1999, providing an alternative to Landsat 5 TM. The ground sampling interval or pixel size for the relevant data is 30 m by 30 m at the equator with one of the enhancements being a panchromatic band that allows for the 30 m visible data to be pan-sharpened to 15 m. The ETM+ repeat cycle was 16 days with a scene size of 170 km by 185 km. As an enhancement on earlier Landsat systems, the TM collected relevant data in six regions of the electromagnetic spectrum whereas the ETM+ system collects information in eight regions of the electromagnetic spectrum.

The three regions in the visible spectrum collected by ETM+ are identical to TM and are known as Band 1 which is in the blue range from 0.45 μm to 0.52 μm, Band 2 which is in the green range 0.52 μm to 0.60 μm, and Band 3 which is in the red range from 0.63 μm to 0.69 μm. The three infrared regions are also identical to TM and are known as Band 4 from 0.76 μm to 0.90 μm in the near-infrared range, Band 5 from 1.55 μm to 1.75 μm which is in the near infrared range, and Band 7 which is in the mid-infrared range from 2.08 μm to 2.35 μm (Table 4.3). These three visible and three infrared bands are entirely comparable to their earlier TM counterparts. Finally, as noted above, a new panchromatic Band 8 was added at 15 m spatial resolution and between 0.52 μm to 0.9 μm was added to ETM+. This new higher-resolution and broader band allows for redefining the spatial resolution of ETM+ Band 2, Band 3, and Band 4 to model a higher resolution than their initial design allows. This process is known as pan-sharpening. Landsat 7 ETM+ follows the WRS-2 reference system (Zanter 2016; Williams 2009; USGS and US Dept. of Interior 2018). The ETM+ path and row for each estuary are included in Table 4.2.

Unfortunately, ETM+ malfunctioned in May 2013 rendering it challenging to use in LULC change analysis. The Scan Line Corrector (SLC), which corrects for the movement of the satellite, failed and was never recovered (Zanter 2016; USGS and US Dept. of Interior 2018). The SLC failure caused intermittent gaps across an ETM+ image which resulted in the central portion of each scene retaining useful information but the rest of the image becoming gradually more void of pixel data as the area of interest was located farther away from the center of a scene. The study site of Muisne Estuary was still mappable with SLC off, as was much of Chone Estuary, but all the other estuaries had to revert to the still operational TM post he SLC malfunction (Table 4.3). Other options aside from reverting to TM for correcting ETM+ included supplementing the ETM+ image with data from another system such as TM, using various algorithmic approaches to fill the gaps, or using another comparable system altogether such as the Japanese ASTER satellite or the French Satellite Pour l'Observation de la Terre (SPOT) 5, SPOT 6, or SPOT 7 systems. Both the ASTER and SPOT systems have comparable radiometric and spatial resolutions to ETM+. Due to the limited period when ETM+ was fully operational with SLC on, the instrument is used sparingly across all estuaries (Table 4.1).

4.3.4 OLI

Landsat 8 OLI launched in 2013 and is still fully operational at the time of publication. The OLI has nine bands of which seven are comparable to those on TM and ETM+ (Zanter 2016; USGS and US Dept. of Interior 2018). Comparable visible and infrared bands exist in both TM and ETM+ when compared to OLI (Table 4.3). For example, OLI Band 2 is the equivalent of TM and ETM+ Band 1, and it operates in the blue range from 0.45 μm to 0.51 μm (Barsi et al. 2014). OLI Band 3 is the equivalent of TM and ETM+ Band 2, and it operates in the green range of 0.53 μm to 0.59 μm (Barsi et al. 2014). OLI Band 4 is the equivalent of TM and ETM+ Band 3, and it operates in the red range from 0.64 μm to 0.67 μm (Barsi et al. 2014). OLI Band 5 operates in the near-infrared range from 0.85 μm to 0.88 μm (Barsi et al. 2014), and this is comparable to Band 4 on TM and ETM+. Band 6 on the OLI is comparable to Band 5 on TM and ETM+ and is now described as shortwave infrared and operates in the short-wavelength infrared (SWIR) range from 1.57 μm to 1.65 μm (Barsi et al. 2014). Finally, OLI Band 7 is comparable to Band 7 on TM and ETM+; it operates in the SWIR range between 2.11 μm and 2.29 μm (Barsi et al. 2014). As with ETM+ and TM, the ground sampling interval or pixel size for the relevant bands listed above are 30 m by 30 m at the equator. The OLI repeat cycle is 16 days with a scene size of 170 km by 185 km (Zanter 2016; USGS and US Dept. of Interior 2018). OLI retains the panchromatic band, this time between 0.50 μm and 2.68 μm, retaining the ability to model Bands 2, 3, and 4 to higher resolutions than the original data allows.

OLI has two additional bands. A new Band 9 named Cirrus that allows for the monitoring of atmospheric pollution or water vapor that may alter image quality (Zanter 2016; USGS and US Dept. of Interior 2018) and the additional Band 1, named Coastal, designed for mapping of shallow water including monitoring phytoplankton, harmful algal blooms, chlorophyll concentration, and suspended sediment. Although these new bands may prove useful for mangrove forest monitoring and will be widely applied in the mangrove forest research community, they are not used in this analysis as this band information is not present for any of the earlier surveys nor most later surveys. OLI is used mostly for the final 2014 surveys (Table 4.1) and is unlikely to be used for future analysis as higher-resolution mangrove forest monitoring systems have since come online. The OLI path and row for each estuary are included in Table 4.2.

4.3.5 ASTER

The ASTER satellite is a joint venture cooperation between Japan's Ministry of Trade and NASA (Abrams et al. 2002). It was launched in December 1999 as part of the Terra (EOS AM-1) platform. Terra follows about 30 min behind ETM+, revisiting sites on an identical schedule as ETM+ every 16 days. The ASTER scene size is 120 km by 150 km. ASTER is essentially three different instrumentations operating at three different resolutions (Abrams et al. 2002).

The first instrument on ASTER is known as the Visible and Near-Infrared scanner (VNIR) and operates at a spatial resolution of 15 m at the equator (Abrams et al. 2002). As the name implies, it captures the visible and near-infrared portions of the electromagnetic spectrum. The four ASTER VNIR bands are referenced as Band 1, Band 2, Band 3 N, and Band 3B. Band 3B on ASTER is a backward facing and time-delayed almost identical capture of band 3 N and allows for digital elevation model (DEM) creation; it is not utilized in this analysis.

The VNIR bands 1, 2 and 3 N are comparable to Landsat TM and ETM+ Bands 2, 3, and 4 or OLI Bands 3, 4, and 5 respectively (Table 4.3). When ETM+ is pan-sharpened, the comparable ETM+ and ASTER VNIR band resolutions would be identical. When ETM+ is not pan-sharpened four ASTER VNIR pixels would fit into each TM, ETM+, or OLI pixel. It should be noted that ASTER does not contain an equivalent to Band 1 which exists in the blue range of the visible spectrum in TM and ETM+, and as Band 2 in OLI. The VNIR sub-instrument on Aster is still operational as of writing (Fig. 4.4). The VNIR sub-instrument on Aster is still operational as of writing (Fig. 4.4).

The second instrument contained in ASTER captures data in the shortwave infrared portion of the spectrum and has six distinct bands with a spatial resolution of 30 m at the equator (Abrams et al. 2002). It is known as the SWIR scanner. The SWIR bands are referred to as Band 4, Band 5, Band 6, Band 7, Band 8, and Band 9. The third portion of ASTER captures data in the thermal infrared (TIR) part of the spectrum at 90 m resolution and is not used in this analysis. SWIR Band 4 is equivalent to TM and ETM+ Band 5 and OLI Band 6 (Table 4.3). SWIR Bands 5, 6, 7, and 8 are smaller radiometric slices of TM and ETM+ Band 7 and smaller radiometric slices of OLI Bands 5, 6, and 7 (Table 4.3). The SWIR portion of ASTER has the same pixel resolution as the equivalent bands in TM, ETM+, and OLI. The SWIR data has been deemed as unusable since April 2018 due to instrument anomalies (Fig. 4.4).

ASTER was not free at the time of analysis, but since 2016, much of the data has since been made available for free internet download via NASA and the USGS to all users globally. ASTER has no path row WRS grid such as those provided by the USGS for TM, ETM+, and OLI downloads and each scene must be retrieved by searching the location and requesting all ASTER scenes.

4.3.6 RapidEye

RapidEye is the only fully commercial satellite system utilized in this study and is only used in Chone Estuary (Table 4.1). Chone is not only oversampled for time (Table 4.1) but is also over-sampled throughout space utilizing RapidEye. The 5 m resolution surveys in Chone Estuary depict more granular LULC information than in other estuaries, allowing for sub-Landsat and sub-ASTER resolution changes to

Table 4.4 RapidEye band information and landsat equivalents

RapidEye	BAND	NAME	μm	TM	ETM+	OLI
Visible	1	Blue	0.440–0.510	1	1	2
	2	Green	0.520–0.590	2	2	3
	3	Red	0.630–0.685	3	3	4
InfraRed	4	Red-edge	0.690–0.730	NA	NA	NA
	5	Near-infrared	0.760–0.850	4	4	5

be documented. RapidEye itself is a constellation of five independent sensors providing over 4,000,000 km^2 of imagery daily and revisiting each site at nadir every 5 days. The RapidEye swath is only 77 km due to the high-resolution of the data achieved by the lower orbit of the system. For this reason, each estuary requires multiple scenes to cover the entire study area. The RapidEye data were ordered with five spectral bands present to best match the MSS, TM, ETM+, OLI, and ASTER visible and near-infrared bands. The spectral bands, and corresponding Landsat and ASTER bands are documented in Table 4.4 (Abrams et al. 2002). Nine RapidEye pixels fit inside of each ASTER pixel, and thirty-six inside TM, ET+, and OLI.

4.4 Mangrove Forest and Shrimp Farm Classification

Extracting comparable LULC from the topographic maps, air photos, and numerous satellite system for each of the years and each of the estuaries in Table 4.1 utilized a suite of automated, manual, and blended approaches.

4.4.1 Topographic Map and Air-Photo Data Processing

For Ecuadorian topographic maps, mangrove forests were delineated for the pre-aquaculture period by heads-up digitizing from the georeferenced topographic maps and additionally in Chone Estuary via manual visual interpretation of the aerial photographs. On the topographic maps, areas with mangrove forest land cover were denoted by a distinct symbol and boundary and were labeled as mangrove forest (Fig. 4.5). The mangrove forest polygons derived from topographic maps were then stored as polygonal features in GIS vector format. Figure 4.5 demonstrates the extraction of mangrove forests from a topographic map in the center of Chone Estuary. The yellow polygons are the extracted data. Note that these polygons are then verified and adjusted based on the high-resolution air-photos. These mangrove forest polygons form the base mangrove forest delineation in each estuary for which topographic maps were available.

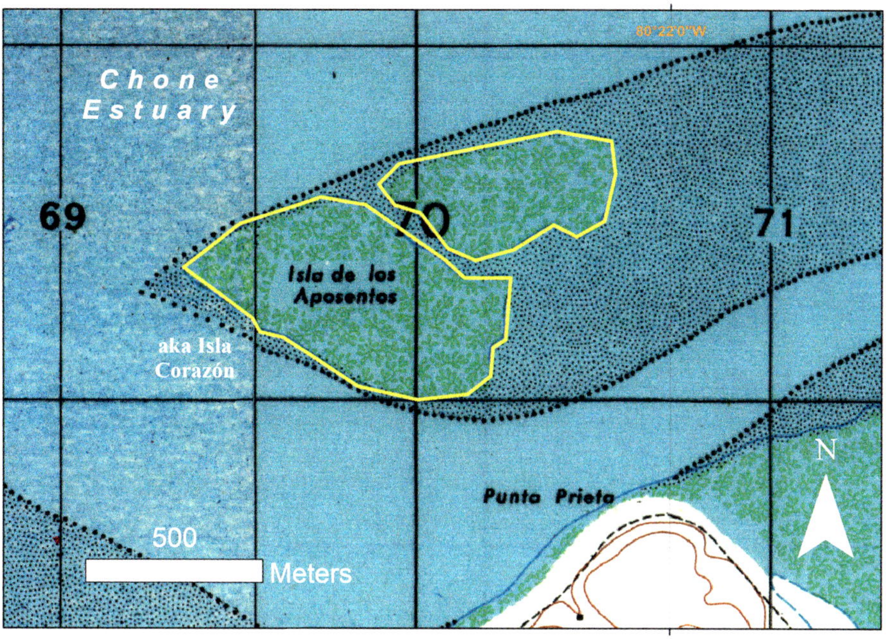

Fig. 4.5 Mangrove forest digitized from a topographic map

Regions of El Oro province have a handful of aquaculture ponds visible in the initial topographic map survey. These ponds are manually digitized in the same manner as the mangrove forests in Fig. 4.5. These areas of shrimp farms in the estuary are denoted with dike markings on earlier maps and shrimp farm terminology is used to identify shrimp ponds on later maps. We verified in the field that these ponds had always been shrimp aquaculture ponds and that they existed at this early time. The shrimp farms from topographic maps, verified with air-photos when possible, were stored as polygonal features in GIS vector format. These shrimp farm polygons form the base shrimp farm delineation in each area where they are used.

Once all shrimp ponds are extracted, then the remaining non-attributed estuarine study area space is categorized merely as *other* LULC. That is, if it is not shrimp farm and it is not mangrove forest; then is labeled as other. The other category is greater than 99% water with some small areas of mud, saltpans, or limited terrestrial landcover. The other areas obtained from topographic maps, verified with air-photos when possible, were stored as polygonal features in GIS vector format. These other landcover polygons form the base non-shrimp farm and non-mangrove forest delineation in each area where they are used.

The final step of mangrove forest, shrimp farm, and other landcover extraction from the early topographic maps and air-photos is to create a topologically coherent singular vector polygon feature class. The integrity of the polygonal fature class is

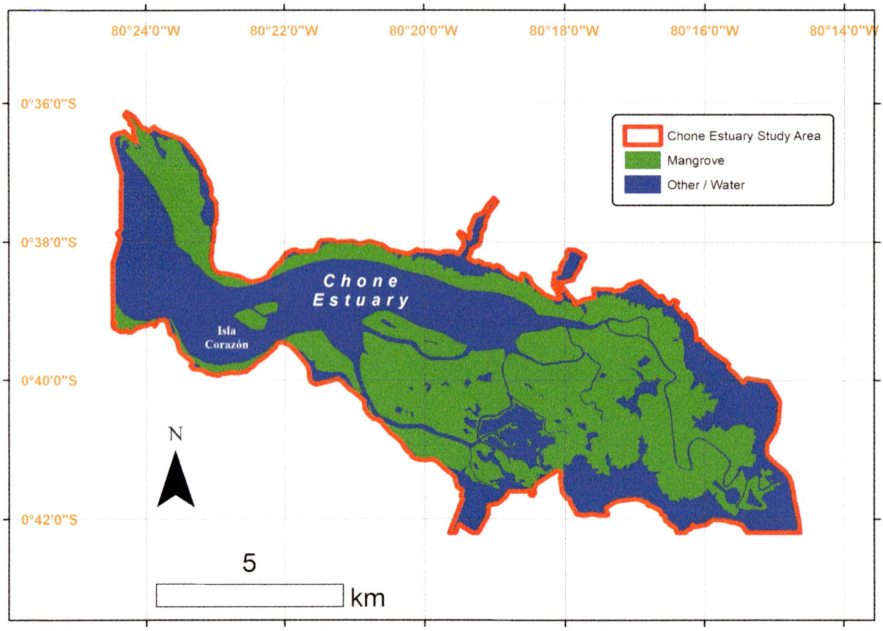

Fig. 4.6 The final raster product for Chone in 1968 derived from topographic maps and air-photos

achieved by building a topology from the study area polygon, the mangrove forests polygons, the shrimp farm polygons, and the other landcover polygons. The topology rules included no overlapping polygons, no gaps between polygons, all landcover polygons must be contained within the estuary study area polygon and a host of other rules such as no exceptionally small polygons. This topology rules process allowed for items such as slivers, gaps, overlap, and null polygons to be identified and repaired.

The final vector feature class for each study site derived from the topographic map and air photos is a single topologically correct feature class that exactly matches the study area extent with three potential classifications for each polygon. The three classes are mangrove forest, shrimp farm, and other LULC (Fig. 4.6). The feature class was then exported to raster format at 1 m resolution for LULC change analysis. Figure 4.6 is the completed output example for Chone Estuary for 1968 LULC data digitized from 1:25,000 topographic maps, supplemented with 1 m air-photos following the process outlined above. The green area is mangrove forests, the blue area is other which is mostly comprised of water, and in this example, as in all estuaries aside from El Oro, the earliest topographic maps and air-photos had no shrimp farms present.

After the initial dataset (initial and 2nd for Chone Estuary) was compiled via the above methods for estuaries that have topographic maps or air-photos listed as their input data source in Table 4.1, the satellite data was processed for each estuary study area.

4.4.2 Satellite Data Processing

All remotely sensed-data was ordered at processing level-2 when possible including TM and ETM+ data which constitute the majority of the surveys conducted (Table 4.1). Level-2 data attempts to recreate the reflectance at the surface at the time of collection. As per NASA, TM and ETM+ surface reflectance data are generated from the Landsat Ecosystem Disturbance Adaptive Processing System, a specialized software initially developed by Goddard Space Flight Center and the University of Maryland (Masek et al. 2006). Level-2 science data account for water vapor, ozone, geopotential height, aerosol optical thickness, and elevation; as well as using radiative transfer to generate top of atmosphere (TOA) reflectance, surface reflectance, TOA brightness, and TOA temperature, while accounting for clouds, cloud shadow adjacent clouds, land, and water (Irish 2000; Chander and Markham 2003; USGS 2019). Landsat Surface Reflectance Level-2 science products are built and intended to support land surface change studies (USGS 2019). A full description of preprocessing of TM and ETM+ data can be found in the associated user's handbooks (Irish 2000; Chander and Markham 2003; USGS 2019). *"Landsat 8 OLI surface reflectance data are generated from the Landsat Surface Reflectance Code (LaSRC), which makes use of the coastal aerosol band to perform aerosol inversion tests, uses auxiliary climate data from MODIS and uses a unique radiative transfer model"* (USGS 2018). A full definition of the preprocessing is included in the most-recent Landsat 8 users handbook (USGS 2018). ASTER data is also processed to level-2. The full description of the processing can be found at NASA et al. (2018). RapidEye data is calibrated to radiance at the sensor, in a product known as RapidEye Orth Level 3A which orthorectifies tile products with radiometric, geometric and terrain corrections applied (Brunn et al. 2016). The small slice of MSS data utilized for a portion of Isla Puná estuary for 1973 uses just digital number values as input.

The satellite delineation of mangrove forests is conducted using unsupervised classification supplemented with normalized difference vegetation index (NDVI), and with the prior period's mangrove forest, shrimp farm, and other landcover classification information. Each input satellite image was limited to each estuarine study area; this was done to limit the number of clusters returned by the unsupervised classification algorithm. Additionally, bands not applicable for this analysis, such as the panchromatic band or the cirrus band were removed from the input image to speed up the classification process and limit the potential of spurious results. The final classification was based on all visual bands and all infrared bands present within the delineated estuarine study area only. When estuaries required more than one scene, the scenes were merged if the collection occurred on the same day, and they were processed as individual scenes if the collection occurred on different days.

Once all data preprocessing is complete, the next step is to extract the relevant LULC categories. The IsoClsuter algorithm was used to obtain landcover signature files for each landcover type inside the estuary. Maximum likelihood classification

(MLC) combined the signature files and important satellite images into clusters based on the information in the signature fields. The combination of IsoCluster and MLC is akin to the IsoData algorithm (Ball and Hall 1965; Richards and Richards 1999). IsoCluster uses an iterative optimized clustering procedure, also known as a migrating means technique. The algorithm divides all pixels into n number of groups in the multidimensional space of all input bands. During each iteration, all the samples are assigned to existing cluster centers and then new means are calculated, the first means established are entirely arbitrary. Numerous reruns of the process are used to reveal the optimum classes, usually two in the case of estuaries dominated by mangrove forests (mangrove forest or not mangrove forest) and more classes in the case of estuaries with heavily fragmented or limited mangrove forests. The outputs are signature files for each identified cluster.

Once signature files are obtained, the next step is to perform a MLC. MLC is based on two principles: the pixels in each pixel sample in the multidimensional space approximating a normal distribution and Bayesian logic (Jensen 2005). The MLC approach considers the variances and covariances of the class signature when assigning a pixel in the image to one of the classes in the signature file, with the assumption that a class sample can be characterized by a mean vector and covariance matrix when the class sample approximates normal (ESRI 2018). Using a probability function, each pixel is then assigned to the class to which it most likely belongs.

The output from the IsoCluster process is a single class, or classes, of mangrove forest and one or more classes of other LULC. The mangrove forest class is then vectorized into a polygon. In this assessment, we are interested in lost mangrove forest or new mangrove forest, so the previous survey's data acts as an input into the vectorization process. For example, if the prior year's mangrove forest has been entirely cleared, that polygon would be deleted. If the prior year's mangrove forest had been somewhat cleared, then the polygon would be adjusted. If any new mangrove forest appeared, then a new polygon would be added. We additionally overlaid an NDVI map of the latest year as this proved useful in detecting the water and mangrove forest boundary. This process is sequential and uses a preponderance of evidence approach. For example, a handful of pixels would not cause a prior year's forest to expand or to contract. There must be substantial evidence of new mangrove forest or cleared mangrove forest for adjustments to be made to the prior survey's polygons. In this analysis, we are interested in mangrove forest change and not the addition or subtraction of very small clusters of mangrove trees. Clear and demonstrable substantial LULC change from the prior year's polygons need to be exhibited for the polygons to be deleted, expanded, contracted, reshaped, or new polygons added; and if doubt remains, the prior year's classification remains in place.

Despite using numerous differing methods, including supplemental data, we failed to extract all shrimp farms in any estuary via an automated process. We could achieve success rates of approximately 60% or 70% but never close to 100%. The reason for the automated shrimp farm extraction failure is threefold. First and most importantly, the water in growout ponds when they are first filled, and for a few days or weeks after they are first filled, has the same spectral signature as the estuary water. Second, the farms contain growout ponds but also contain other features such as structures, dikes, mud, grass, and other vegetation. These were difficult to separate from other terrestrial landcover within the study areas. Third, empty ponds have mud bottoms which have the same spectral signature as the other wet mud bottoms in the images. To reliably extract shrimp farms, an object approach using software such as Ecogniton is required that could work on a rule-based and shape-based classification as opposed to a pixel classification, or at least to work in conjunction with a pixel-classifier. As this approach is a research frontier with no guarantee of success, we did not attempt this on more than a few test areas, although the tests did prove promising.

On satellite imagery, shrimp farms display as distinctive polygons with straight edges, generally occurring in clusters within the estuary (Fig. 2.3). The clusters of ponds are many hectares in size and large enough to be viewed with the spatial resolution of 30 m from Landsat products. Pan-sharpening the images when possible leads to even more accurate observation of pond boundaries. Indeed, the average size of the growout ponds is approximately 10 ha, and very few are under 5 ha (Fig. 2.3). Assuming the generally rectangular pattern of shrimp farms and an average size of 10 ha (Fig. 2.3), approximately 111 Landsat, 111 ASTER, and 4000 RapidEye pixels reside within each growout pond. These size and shape attributes of shrimp farms, combined with the fact that nothing else of this regular shape and size exists in an estuary, makes it possible to digitize the shrimp ponds manually. Manual digitization was a time-consuming task that took many tens-of-thousands of person-hours across multiple people, but the output product is an accurate and precise series of vector polygon shrimp farms for each period analyzed.

When digitizing shrimp farms, care needs to be taken not to confuse growout ponds with salt-ponds, rice fields, sewage treatment plants, or power-generation cooling ponds. Field verification can resolve potential conflicts, but each of these non-shrimp farm landcover has distinct features that often rule them out from being shrimp farms without the need for field verification. For example, cooling ponds are associated with power generation, and large power plants are easy to locate in the imagery. Wastewater treatment plants are not inside estuaries, and even when on the edge of an estuary such features have a distinct pie-slices shape and unique associated infrastructure. Likely the most challenging LULC to distinguish from shrimp ponds are artificial salt-ponds. As opposed to shrimp growout ponds, salt-ponds are generally located in areas of low humidity and rainfall, are often highly irregular in shape following natural landscape features, are far larger than shrimp farms (100 ha), and have periods when salt is visible on the bottom of the pond.

Once the mangrove forests and the shrimp farm polygon creation is complete for a survey year, the two sets of features are merged. Next, any estuarine study area not classified as mangrove forest or shrimp farm in an estuary is given the other classification. Again, other is mostly surface water. The topology rules again included no overlapping polygons, no gaps between polygons, all landcover polygons must be contained within the estuary polygon and a host of other rules such as eliminating exceptionally small polygons under 1 ha. The final feature class was then exported to raster format at 1 m for LULC change analysis. Figure 4.7 is the completed out-

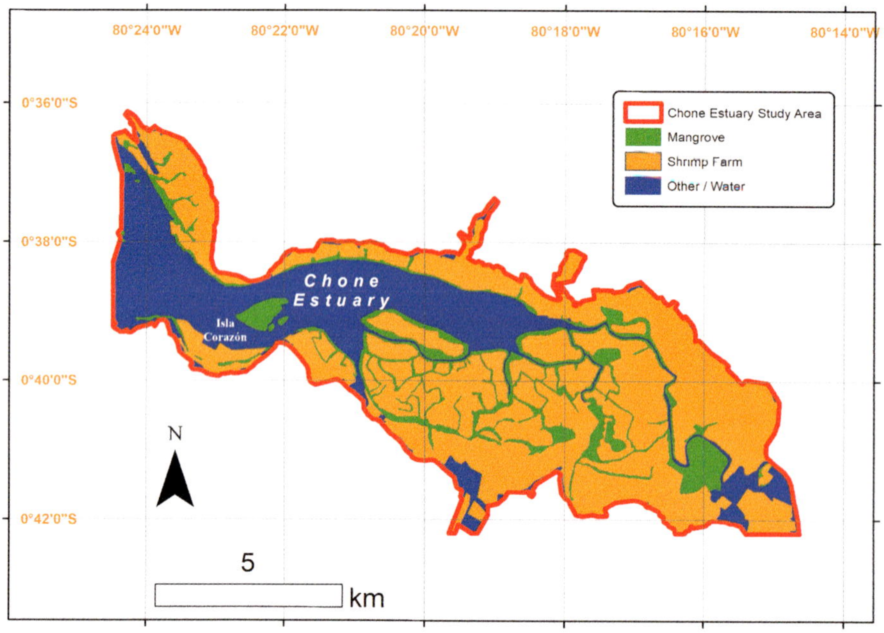

Fig. 4.7 The final raster product for Chone in 1998 derived from TM data

put example for Chone Estuary for 1998 LULC data derived via the IsoCluster process from 1998 Landsat data. The green areas are mangrove forests, the blue areas are other classification, which is mostly comprised of water, and the orange area are shrimp farms. This above method was applied to all survey years that used MSS, TM, ETM+, OLI, ASTER, and RapidEye remotely sensed data (Table 4.1).

4.5 LULC Change Analysis from the Advent of Commercial Shrimp Aquaculture to 2014

Once LULC classifications for all years are stored in raster format and pixel alignment ensured, the process of assessing LULC change from year to year is quite straightforward using basic map algebra. The process can be done using numerous off-the-shelf software tools or free and open-source software. Change has occurred when a pixel's LULC classification changes from one survey period to the next. For example, if a pixel was mangrove forest in the 1970s classification and it converted to shrimp farm in the 1980s, then LULC change has occurred. The most common name for the GIS process that detects LULC change in this manner is a cross-tabulation. A simple way to conduct this form of pixel-to-pixel change detection with only three LULC classes is to merely subtract the period of observation from

Fig. 4.8 Detecting when LULC change has occurred

$$T^{-1} - T^{0} \neq 0$$

T^0 = period of observation and T^{-1} = the prior survey period.

Table 4.5 Obtaining the direction of LULC change

LULC		LULC			
T^0		T^{-1}		Output	LULC change occurred
1	–	1	=	0	Mangrove forest – no change
3	–	3	=	0	Shrimp farm – no change
7	–	7	=	0	Other LULC – no change
1	–	3	=	−2	Mangrove forest > shrimp farm
3	–	7	=	−4	Shrimp farm > other LULC
7	–	1	=	6	Other LULC > mangrove Forest
1	–	7	=	−6	Mangrove forest > other LULC
3	–	1	=	2	Shrimp farm > mangrove Forest
7	–	3	=	4	Other LULC > shrimp farm

the prior period at the pixel level, and if the outcome is anything other than zero, then LULC change has occurred between the two periods (Fig. 4.8).

It is not only the presence of change that is important but the direction of change. That is, it is important to distinguish between a conversion from mangrove forest to shrimp farm and a conversion from shrimp farm to the mangrove forest; just knowing change has occurred is not enough. By using map algebra, the direction of change can be obtained by using a unique class value for each LULC type, and also produces unique outcomes when subtracted. In this example, we use a class of 1 for mangrove forest, a class of 3 for shrimp farms, and a class of 7 for the other LULC (Table 4.5). By using the unique values presented, all subtraction outputs are unique and the direction of LULC from one period to the next or between two survey periods can be obtained. The suite of six possible LULC transitions are listed in Table 4.5, as well as the three values that equal no LULC change. In this manner, the change direction at each pixel is determined.

It is not only the presence of change and the direction of change that is important but the magnitude of change. That is, how many pixels have transitioned from one LULC to another. Quantifying the magnitude of change is achieved by merely summing the number of pixels for each LULC change in Table 4.5. As the raster pixel size is 1 m on the x-axis and 1 m on the y-axis, the sum of the changed pixels in the amount of change is square meters. The sum of the pixels (Table 4.6) measures the magnitude of a LULC change at the estuarine level in square meters. In the hypothetical example in Table 4.6, 100,000 m^2 of mangrove forest remained mangrove forest, 12,000 m^2 of mangrove forest converted to shrimp farm, and 2000 m^2 of mangrove forest was converted to other LULC. Using this system, an analyst can introduce any additional polygon, such as protected areas and sum the output values

Table 4.6 Obtaining the direction and magnitude of LULC change

LULC T^{-1}		LULC T^0		OUTPUT	LULC Change Occurred	SUM m^2
1	–	1	=	0	Mangrove forest no change	100,000
3	–	3	=	0	Shrimp farm no change	80,000
7	–	7	=	0	Other LULC no change	60,000
1	–	3	=	−2	Mangrove forest > Shrimp farm	12,000
3	–	7	=	−4	Shrimp farm > Other LULC	6000
7	–	1	=	6	Other LULC > Mangrove Forest	1000
1	–	7	=	−6	Mangrove forest > Other LULC	2000
3	–	1	=	2	Shrimp farm > Mangrove Forest	500
7	–	3	=	4	Other LULC > Shrimp farm	250

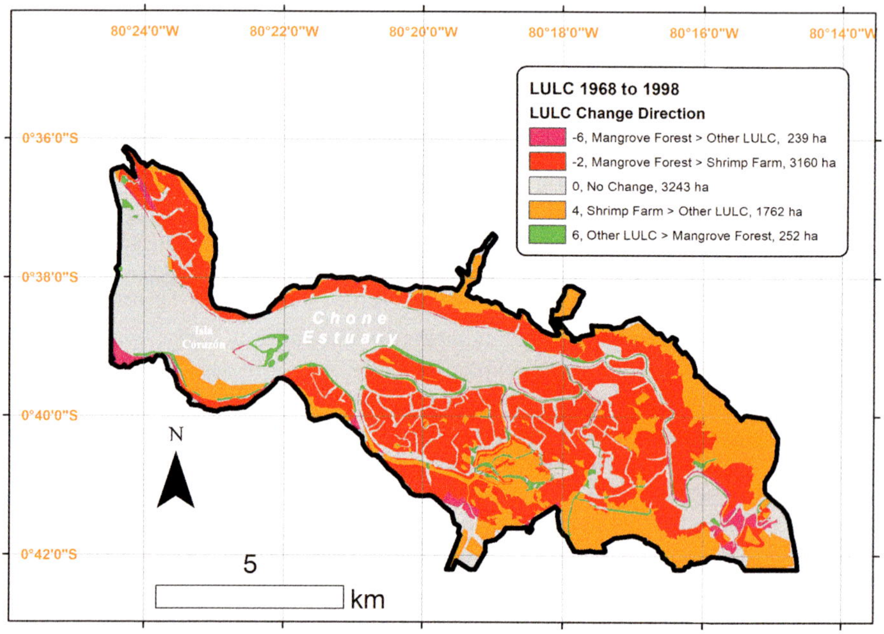

Fig. 4.9 Obtaining the direction and magnitude of LULC change at the estuary and pixel level

within that polygon, and the output would be the magnitude of each possible LULC change in that selected area.

A completed LULC analysis from one survey period to another in one estuary is represented in Fig. 4.9. In this example from the actual data, we examine the 1998 survey in Chone Estuary (Fig. 4.7) and subtract this from the 1968 survey (Fig. 4.6). The observation period is the most recent LULC survey (Fig. 4.8, T^0) which in this case is 1998; we subtract 1998 from the prior period LULC survey (Fig. 4.8, T^{-1}) which in this demonstration is 1968. Therefore, this is 1968 minus 1998 at the pixel level. By doing this, we obtain all the LULC change that occurred and the magnitude of this change (Fig. 4.9).

In Fig. 4.9, the red value represents all area that was mangrove forest in 1968 but is now shrimp farm in 1998. The value represents the change from mangrove forest to shrimp aquaculture, −2 in Table 4.6, and hence in Fig. 4.9. The magnitude of this change is 3160 ha. That is, between 1968 and 1998, 3160 ha of 1968 mangrove forest is 1998 shrimp farm. More than the entire estuary levels of change, the LULC change process reveals the location of the mangrove stands lost; the majority of mangrove stands that transitioned were in the interior portions of the estuary. In Fig. 4.8, the purple value of −6 represents areas that were mangrove forest in 1968 but have transitioned to another LULC besides shrimp farms by 1998. In Fig. 4.8, the grey value of 0 represents areas that have the same landcover in 1998 as they did in 1969. In Fig. 4.8, the orange value of 4 represents areas that were mangrove forest in 1968 but have transitioned to another LULC besides shrimp farms by 1998. Finally, in Fig. 4.8, the green value of 6 represents areas that were other LULC in 1968 but have transitioned to mangrove forest by 1998. The values 2 and -4 (Table 4.6) are not present in Fig. 4.9 because no shrimp farms are present in 1968 (Fig. 4.6) so the transition value of 2 and the transition value of −4 are not possible; with no shrimp farms present in 1968, they could not transition to mangrove forest or other LULC respectively. These two transitions are possible in later years when shrimp farms are present.

The map algebra LULC change detection was conducted for every sequential pair of surveys in every estuary, at every decadal period in every estuary, and using the first and last survey in every estuary. By conducting this analysis in this manner, the change is detected at the most granular scale of survey to survey, and at consistent and comparable decadal intervals, in addition to across the whole of the analysis time frame and spaces.

4.6 Remote Sensing Summary

Remote sensing of mangrove forests has many unique challenges and is considered one of the most difficult forest remote sensing tasks (Heumann 2011; Kuenzer et al. 2011). Therefore, ground-truthing fieldwork often remains a part of the remote sensing of mangrove forest process to either determine the accuracy of the mangrove delineation or contribute to the mangrove delineation (Hamilton et al. 2018). This paper follows two of the methodologies for the remote sensing of mangrove forests outlined in Hamilton et al. (2018). These are 22.6.1, Photography Data and Methods, for some of the earliest surveys with map or photo input data and 22.6.2, High and Medium Resolution Satellite-Based Optical Imagery, for most of the later surveys. These methods are supplemented with fieldwork and limited use of section 22.6.3, Very High-Resolution Satellite-Based Optical Imagery (VHR) (Hamilton et al. 2018) for one final survey in Chone Estuary.

The remote sensing method presented allows for the delineation of mangrove forests, shrimp farms, and other LULC across all of Ecuador's estuaries from the initial period of commercial shrimp farm arrival in each estuary to 2014. It allows for the three LULC-derived classifications to be comparable across all the estuarine space and all the time-slices provide by the satellite imagery and air-photos. The granularity allows for analysis at the sub-estuarine level such as sections of the estuary a few hectares in size, and the parish level, canton level, provincial level, and due to the sites covering greater than 99.9% of all historic mangrove forest cover, also at the national level. The method allows not only for monitoring the LULC change occurring in each estuary, but most importantly, allows for the actual LULC displacement types to be ascertained and quantified.

One final methodological consideration for the past surveys is to allow for conducting future surveys with known instruments, planned instruments, and yet unknown and unplanned instruments. For example, a suite of surveys will be conducted in the mid-2020s for all estuaries and in 2019 for Chone Estuary. The most likely instrument to be used for these future surveys is the European Space Agency (ESA) instrument suite known as Sentinel-2, should it remain operational. These future data will increase the spatial resolution to 10 m and include as many as 13 spectral bands. Such systems are approaching the spatial resolution of commercial satellite systems and have increased radiometric resolution. The methodology applied will allow for integration of data in the future from such instruments as Sentinel-2, whereas an instrument-specific, or even a resolution-specific approach, would have nullified the potential for continuing this long-term analysis.

As technology continues to improve and costs drop, it may soon be possible to collect data at the sub-meter resolution at little cost for most Ecuadorian estuaries. Traditional aerial photography, with suitable sensors, is both costly and requires substantial infrastructure on the ground. The potential for unmanned aerial vehicle (UAV) imagery technology to map mangrove forests is a research frontier that shows excellent promise. Indeed, tests are underway to map mangrove forests, and in some cases, to compare these data against EOS-derived imagery (Tian et al. 2017; Ruwaimana et al. 2018; Hamilton and Castellanos-Galindo 2018) and integrate field collected data (Otero et al. 2018). Such UAV technology may allow for more surveys to be conducted than before, and these surveys will be at a higher granularity and available on an as-needed basis. Again, the methodology used in this manuscript will allow for integration of mangrove forest surveys should they transition from EOS to UAV imagery.

The method is presented in a step-by-step repeatable format, with links provided to each of the input and output datasets required to conduct the remote sensing analysis. The remote sensing analysis, going back to 1968, at this sampling interval, at this granularity is likely one of the longest time series analysis ever undertaken for an entire nation for any LULC-type globally.

Recommended Further Reading

Remote Sensing and Digital Image Processing Including the Unsupervised Classification Process

Jensen (2005)
Jensen (2007)

Landsat Guides

USGS (2018)
Irish (2000)
Chander and Markham (2003)

ASTER Guides

Abrams et al. (2002)

Remote Sensing of Mangrove Forests

Hamilton et al. (2018)
Heumann (2011)
Kuenzer et al. (2011)

Map Algebra

Otero et al. (2018)
ESRI (2016)
QGIS Development Team (2016)

Data Availability

All remotely sensed MSS, TM, ETM+, OLI, and ASTER data, aside from the commercial RapidEye data, are available from GLOVIS and Earth Explorer using the parameters provided.

The WRS-1 and WRS-2 grids are available in shapefile format from the USGS at https://Landsat.usgs.gov/pathrow-shapefiles.

The fishnet created by the authors is available at https://doi.org/10.7910/DVN/DPCAST

The study areas are available at https://doi.org/10.7910/DVN/7U8OHW

The air-photos used in this analysis are available at https://doi.org/10.7910/DVN/ZMUC9P

The topographic maps used in this analysis are available at https://doi.org/10.7910/DVN/21FMNQ

The topographic map index is available at https://doi.org/10.7910/DVN/NG6HVZ

A supplementary set of air photographs are available at https://doi.org/10.7910/DVN/ZMUC9P

Cayapas-Mataje Estuary final classified LULC analysis data for each year (all of Table 4.1) are available at https://doi.org/10.7910/DVN/7DB68C

Muisne Estuary final classified LULC analysis data for each year (all of Table 4.1) are available at https://doi.org/10.7910/DVN/7DB68C

Cojimies Estuary final classified LULC analysis data for each year (all of Table 4.1) are available at https://doi.org/10.7910/DVN/7DB68C

Muisne Estuary final classified LULC analysis data for each year (all of Table 4.1) are available at https://doi.org/10.7910/DVN/7DB68C

Chone Estuary final classified LULC analysis data for each year (all of Table 4.1) are available at https://doi.org/10.7910/DVN/7DB68C

Isla Puná North Estuary classified LULC analysis data for each year (all of Table 4.1) are available at https://doi.org/10.7910/DVN/7DB68C

Isla Puná Estuary classified LULC analysis data for each year (all of Table 4.1) are available at https://doi.org/10.7910/DVN/7DB68C

Guayas Estuary classified LULC analysis data for each year (all of Table 4.1) are available at https://doi.org/10.7910/DVN/7DB68C

El Oro Estuary classified LULC analysis data for each year (all of Table 4.1) are available at https://doi.org/10.7910/DVN/7DB68C

References

Abrams M, Hook S, Ramachandran B (2002) ASTER user handbook, version 2. Jet propulsion laboratory 4800:135

Ball GH, Hall DJ (1965) ISODATA, a novel method of data analysis and pattern classification. Stanford Research Institute, Menlo Park

Barsi JA, Lee K, Kvaran G, Markham BL, Pedelty JA (2014) The spectral response of the Landsat-8 operational land imager. Remote Sens 6(10):10232–10251

Brunn A, Bahloul S, Hoffmann D, Anderson C (2016) Recent progress in in-flight radiometric calibration and validation of the RapidEye constellation of 5 multispectral remote SENSING satellites. In, Cham, 2016. Image and Video Technology – PSIVT 2015 Workshops Springer International Publishing, pp 273–284

Chander G, Markham B (2003) Revised Landsat-5 TM radiometric calibration procedures and postcalibration dynamic ranges. IEEE Trans Geosci Remote Sens 41(11):2674–2677

ESRI (2016) http://desktop.arcgis.com/en/arcmap/10.3/tools/spatial-analyst-toolbox/raster-calculator.htm. ESRI. http://desktop.arcgis.com/en/arcmap/10.3/tools/spatial-analyst-toolbox/raster-calculator.htm. Accessed 12 Jan 2018

ESRI (2018) How maximum likelihood classification works. ESRI. http://desktop.arcgis.com/en/arcmap/10.3/tools/spatial-analyst-toolbox/how-maximum-likelihood-classification-works.htm. Accessed 12, 2018

Hamilton SE, Castellanos-Galindo GA (2018) UAV derived, 6 cm resolution. GeoTIFF, RGB, Projected, GIS-ready raster dataset of Rio Preguicas Mangrove Stand, Maranhão State, Brazil

Hamilton SE, Castellanos-Galindo GA, Millones-Mayer M, Chen M (2018) Remote sensing of mangrove forests: current techniques and existing databases. In: Makowski C, Finkl CW (eds) Threats to mangrove forests: hazards, vulnerability, and management, vol 25. Springer, Cham, pp 497–520. https://doi.org/10.1007/978-3-319-73016-5_22

Heumann BW (2011) Satellite remote sensing of mangrove forests: recent advances and future opportunities. Prog Phys Geogr 35(1):87–108

Irish RR (2000) Landsat 7 science data users handbook. National Aeronautics and Space Administration, Report 2000, pp 415–430

Jensen JR (2005) Introductory digital image processing: a remote sensing perspective, Prentice Hall series in geographic information science, 3rd edn. Prentice Hall, Upper Saddle River

Jensen JR (2007) Remote sensing of the environment: an earth resource perspective, Prentice Hall series in geographic information science, 2nd edn. Prentice Hall, Upper Saddle River

Kuenzer C, Bluemel A, Gebhardt S, Quoc TV, Dech S (2011) Remote sensing of mangrove ecosystems: a review. Remote Sens 3(5):878–928

Masek JG, Vermote EF, Saleous NE, Wolfe R, Hall FG, Huemmrich KF, Gao F, Kutler J, Lim T-K (2006) A Landsat surface reflectance dataset for North America, 1990-2000. IEEE Geosci Remote Sens Lett 3(1):68–72

NASA, METI, AIST, Japan Spacesystems, U.S./Japan ASTER Science Team (2018) ASTER on-demand L2 surface reflectance VNIR and SWIR crosstalk-corrected, vol 2018. LP DAAC, Washington, DC. https://doi.org/10.5067/ASTER/AST_07XT.003

NIMA (2000) Department of defense world geodetic system 1984: its definition and relationships with local geodetic systems. 3rd, Amendment 1 edn. NIMA, Bethesda

Otero V, Van De Kerchove R, Satyanarayana B, Martínez-Espinosa C, Fisol MAB, Ibrahim MRB, Sulong I, Mohd-Lokman H, Lucas R, Dahdouh-Guebas F (2018) Managing mangrove forests from the sky: forest inventory using field data and unmanned aerial vehicle (UAV) imagery in the Matang Mangrove Forest Reserve, peninsular Malaysia. For Ecol Manag 411:35–45

QGIS Development Team (2016) QGIS user manual: raster calculator

Richards JA, Richards J (1999) Remote sensing digital image analysis, vol 3. Springer, Berlin

Ruwaimana M, Satyanarayana B, Otero V, Muslim AM, Syafiq M, Ibrahim S, Raymaekers D, Koedam N, Dahdouh-Guebas F (2018) The advantages of using drones over space-borne imagery in the mapping of mangrove forests. PLoS One 13(7):e0200288

Tian J, Wang L, Li X, Gong H, Shi C, Zhong R, Liu X (2017) Comparison of UAV and WorldView-2 imagery for mapping leaf area index of mangrove forest. Int J Appl Earth Obs Geoinf 61:22–31

USGS (2018) Landsat 8 surface reflectance code (LASRC) product guide, vol 1. USGS, Sioux Falls

USGS (2019) Landsat surface reflectance Level-2 science products support land surface change studies. USGS. https://landsat.usgs.gov/landsat-surface-reflectance-data-products. Accessed 12 Jan 2018

USGS, US Dept. of Interior (2018) Landsat Missions. USGS. https://landsat.usgs.gov/. Accessed 12 Jan 2018

Williams D (2009) Landsat-7 science data User's handbook. National Aeronautics and Space Administration, p 186

Woodcock CE, Allen R, Anderson M, Belward A, Bindschadler R, Cohen W, Gao F, Goward SN, Helder D, Helmer E (2008) Free access to Landsat imagery. Science 320(5879):1011–1011

Zanter K (2016) Landsat 8 (L8) data users handbook. Landsat Science Official Website Available online: https://landsat usgs gov/landsat-8-l8-data-users-handbook. Accessed 20 Jan 2018)

Chapter 5
Results

Abstract This chapter presents the results of the land use and land cover (LULC) remote sensing analysis as described in Chap. 4. It describes the estuarine LULC change that occurred in each estuary from the arrival of shrimp aquaculture to 2014. The results are presented at the estuary level for each estuary. Section 5.2 moves beyond LULC change at the estuary level and examines LULC displacement within each estuary. It is dedicated to explaining the displacement type that occurred in each estuary. That is, if mangrove forest was lost, when the loss occurred and precisely what LULC replaced the mangrove forest there. Section 5.3 does the same as Sect. 5.1 but with all results aggregated to the national level. Section 5.4 does the same as Sect. 5.2 but with all results aggregated to the national level. Section 5.5 examines other results suggested by the remote sensing analysis at the sub-estuarine level and Sect. 5.6 summarizes the finding of this chapter. The results demonstrate a massive conversion from mangrove forest to shrimp aquaculture across almost all estuaries and that the preferred location for new shrimp farms was former mangrove forest.

Keywords Mangrove loss · Mangrove gain · Land use change · Land cover change · Shrimp farm expansion · Aquaculture expansion

5.1 Landuse and Landcover Change from the Arrival of Shrimp Aquaculture to 2014 at the Estuarine Level

The coastal estuaries' LULC change results are presented below starting at the extreme north of Ecuador on the Colombian border with Cayapas-Mataje Estuary and continuing from north to south through all the coastal estuaries, ending on the extreme southern border of Ecuador at the Peruvian border in the estuaries of El Oro Province (Fig. 3.1).

Electronic Supplementary Material The online version of this chapter (https://doi.org/10.1007/978-3-030-22240-6_5) contains supplementary material, which is available to authorized users.

S. E. Hamilton, *Mangroves and Aquaculture*, Coastal Research Library 33,
https://doi.org/10.1007/978-3-030-22240-6_5

5.1.1 Cayapas-Mataje Estuary

Figure 5.1 depicts the three LULC designations across Cayapas-Mataje Estuary from the period before the arrival of shrimp farms to 2014. In 1986, Cayapas-Mataje Estuary had no shrimp farms present and was dominated by mangrove forests and the other LULC designation which in this estuary is comprised almost entirely of water. In 1986, the estuary had 35,057 ha of mangrove forest, 15,568 ha of other LULC, and no observable shrimp farms (Fig. 5.1). In the central and western portions of Isla de Santa Rosa and on the periphery of the study site boundary, some mangrove forests were mixed with palm vegetation and other terrestrial vegetation,

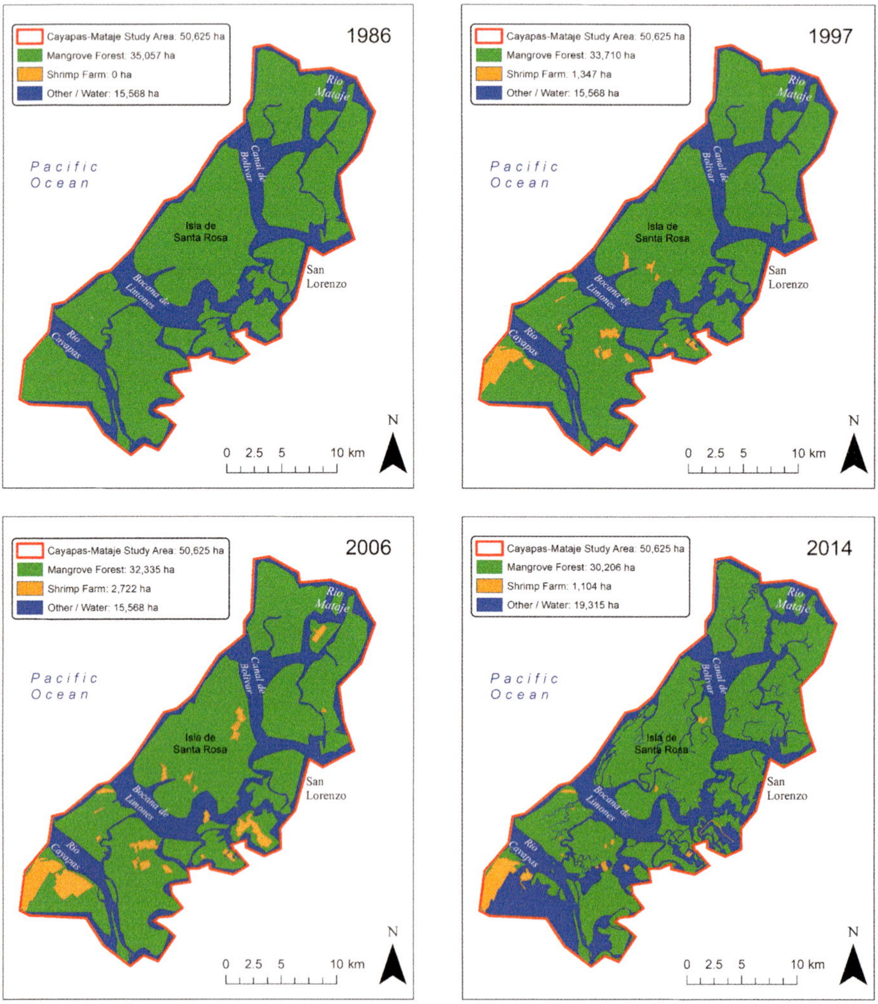

Fig. 5.1 Cayapas-Mataje estuary LULC change from the arrival of shrimp aquaculture to 2014

but the area was dominated by mangrove forest, so the mangrove forest designation remains appropriate for these areas.

By 1997, shrimp farms had appeared within the Cayapas-Mataje Estuary, but the estuary is still dominated by mangrove forests and the other LULC designation (Fig. 5.1). The shrimp farms appeared in a large cluster near La Toya (Figs. 3.2 and 5.1) in the southwest corner of the study area, and to a lesser extent in the central and interior of the estuary. The northern portion of the estuary remained free of shrimp farms. The REMACAN site appears to have been delineated to purposefully avoid this cluster of shrimp farms around La Toya (Figs. 3.2 and 5.1). Indeed, the REMACAN narrows to a few meters wide to avoid the shrimp farms and then expands out to capture more mangrove forest. The 1998 shrimp farm area in the estuary was 1347 ha, but mangrove forest remains dominant, covering over 33,710 ha of the estuary, and the other LULC classification remained unchanged from 1986 at 15,568 ha (Fig. 5.1). Cayapas-Mataje Estuary remained a mangrove forest dominated estuary in 1986, but shrimp farms are starting to take up limited estuarine space.

By 2006, the area of shrimp farms has almost doubled in Cayapas-Mataje Estuary, but the estuary is still dominated by mangrove forests and the other LULC designation (Fig. 5.1). Many of the new shrimp farms appeared in areas adjacent or almost adjacent to the preexisting shrimp farms, but by 2009, shrimp farms have appeared in the northern portion of the estuary near Colombia. The 2006 shrimp farm level in the estuary was 2722 ha, but mangrove forest remains the dominant LULC, covering 33,710 ha of the estuary, and the other LULC classification remains unchanged from 1997 at 15,568 ha (Fig. 5.1). Cayapas-Mataje Estuary remained a mangrove forest dominated estuary in 1986, but shrimp farms appear to be expanding at a rapid rate in the early 2000s.

By 2014, shrimp farms are in retreat in Cayapas-Mataje Estuary, and the estuary is still dominated by mangrove forests and the other LULC designation (Fig. 5.1). Many of the shrimp farms that appeared between 1997 and 2006 are no longer present, and the northern portion of the estuary is once again almost shrimp farm-free as it was in 1986. The 2014 shrimp farm level in the estuary is 1104 ha, which is close to the same amount of shrimp farm coverage as in 1986. Mangrove forest remains the dominant LULC, covering 30,206 ha of the estuary, and the other LULC classification has decreased slightly to 19,315 ha (Fig. 5.1). The gain of the other LULC class and the loss of the mangrove forest between 2006 and 2014 appears to be partially on the terrestrial edge of the estuary. This may be due to improved instrument resolution as opposed to LULC change. The other LULC class gains appear to be at the expense of shrimp farms.

Figure 5.2 shows the percentage of each LULC in Cayapas-Mataje Estuary. As of 1986, mangrove forest was 69% of the LULC present in the estuary and decreased to 60% by the final survey. Mangrove forest has declined in Cayapas-Mataje during the analysis period but not significantly. Shrimp farms, by study design, started at 0% of LULC in 1986 and became as much as 5% of all estuarine land cover by 2006. By 2014, shrimp farm LULC had decreased to 2% of estuarine LULC, below their 1997 LULC area (Fig. 5.2). Despite gaining a foothold in the estuary in the

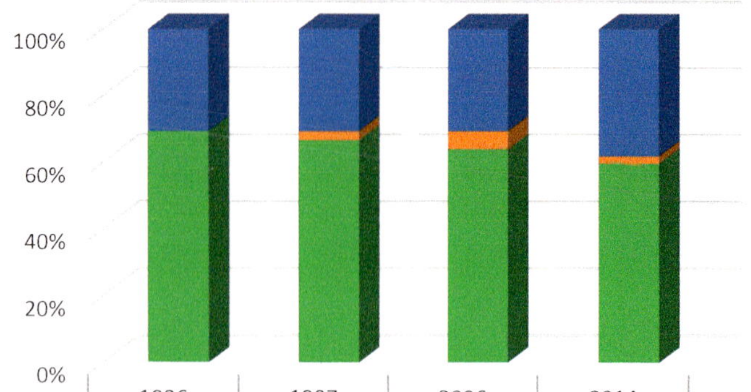

	1986	1997	2006	2014
Other LULC	31%	31%	31%	38%
Shrimp Farms	0%	3%	5%	2%
Mangrove Forest	69%	67%	64%	60%

Fig. 5.2 Cayapas-Mataje estuary LULC change over time

1990s, it appears that shrimp farming did not continue its expansion in the 2010s. A future survey is required to ascertain if the decrease in shrimp farm LULC is an actual trend or merely an anomaly. To help assess the interdecadal LULC changes in Cayapas-Mataje Estuary two additional surveys for 2001 and 2011 are included in the accompanying Dataverse data repository.

5.1.2 Muisne Estuary

Figure 5.3 depicts the three LULC designations present in Muisne Estuary from the period before the arrival of shrimp farms in 1970–2014. In 1970, Muisne Estuary had no shrimp farms present and was dominated by mangrove forests and the other LULC designation, which in this estuary is comprised of water, salt-pans, mudflats, and some limited terrestrial environment. Mangrove forest is the dominant LULC, and the other LULC designation is also substantial. Mangrove forests covered 2947 ha of the estuary with mangrove forests dominating the interior portions of the central and northern parts of the estuary (Fig. 5.3). The other LULC designation, comprised mostly of water, salt-pans, and mudflats occupied 2703 ha of estuarine space. In 1970, Cojimies Estuary appears to have had both large and contiguous mangrove forests within large portions of the estuary. It should be noted that Muisne Estuary is approximately one-tenth of the size of Cayapas-Mataje Estuary.

By 1986, shrimp farms gained a foothold within Muisne Estuary, but the estuary is still dominated by mangrove forests and the other LULC designation (Fig. 5.3). The shrimp farms appeared in three clusters: one in the south of the study area, one on the coastal part of Muisne Island near the town of Muisne, and one near the town

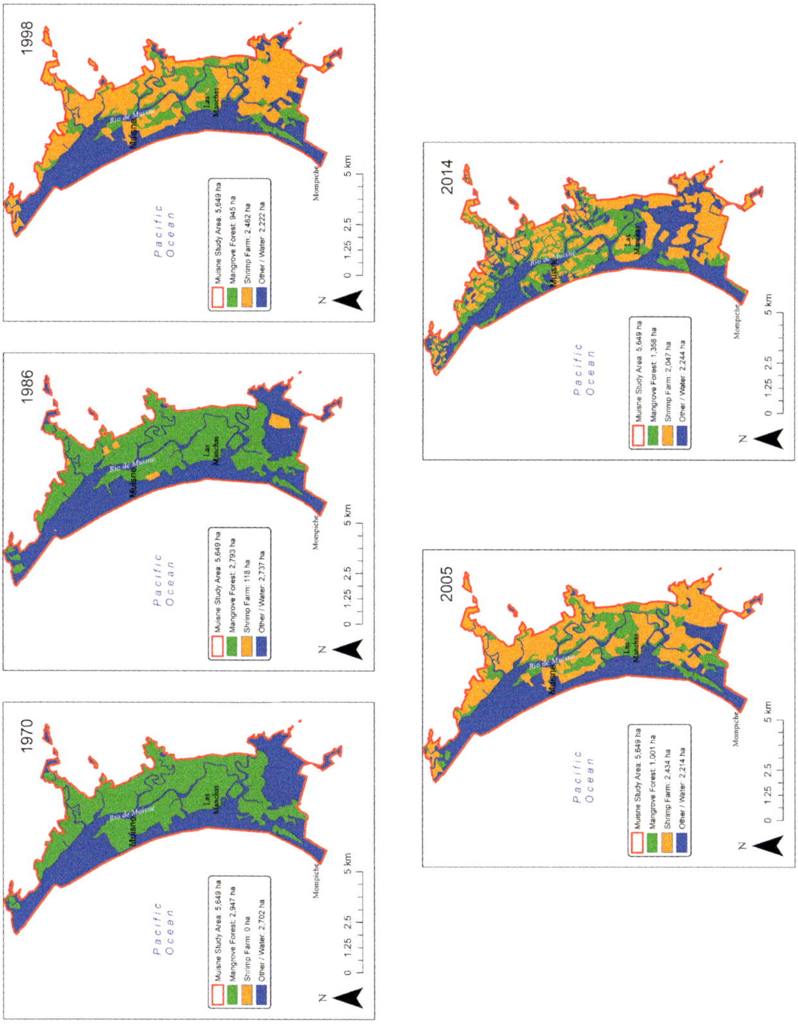

Fig. 5.3 Muisne estuary LULC change from the arrival of shrimp aquaculture to 2014

of Muisne on the mainland. The 1998 shrimp farm level in the estuary was only 118 ha, and the 1986 satellite image appears to have captured the shrimp farm development in a nascent stage. Mangrove forest remains the dominant LULC, covering 2793 ha of the estuary, and the other LULC classification remains approximately stable at 2737 ha (Fig. 5.3). Muisne Estuary remained a mangrove forest dominated estuary in 1986, but shrimp farms are starting to consume some limited portions of the estuary.

By 1998, mangrove forests in Muisne Estuary appear to have undergone substantial clearing. In 1986, 2793 ha of mangrove forests existed in the estuary, but by 1998, this had declined to only 945 ha (Fig. 5.3). Muisne Estuary lost two-thirds of its mangrove forest in these 12 years. Conversely, by 1998, shrimp farms have expanded to a remarkable level in Muisne Estuary (Fig. 5.3). Shrimp farms consumed only 118 ha of estuary space in 1986, but a little over one decade later, they had increased twenty-fourfold to 2842 ha (Fig. 5.3). At this time, shrimp farms are the dominant LULC of the estuary, consuming more estuary space than either mangrove forest or the other LULC classification. It would now be correct to state that some limited mangrove forests exist in a shrimp farm environment as opposed to the more traditional designation of shrimp farming occurring in the mangrove forest environment. The other LULC class appears to have undergone some minimal losses between 1986 and 1998, losing 515 ha (Fig. 5.3) of their LULC.

Between 1998 and 2005, the LULC of Muisne Estuary essentially remained stable (Fig. 5.3). Neither mangrove forest cover, shrimp farm cover, not the other LULC class have deviated in these 7 years by more than 50 ha.

Between 2005 and 2014, the mangrove forests of Muisne Estuary appear to have expanded in numerous locations across the estuary (Fig. 5.3). The gain of 357 ha constitutes approximately one-third of the lowest mangrove forest cover observed in 1998. New forests have appeared north of the town of Muisne, and there appears to be an extension of preexisting forests in the south just north of Mompiche. To put this recovery in perspective, mangrove forests in Muisne Estuary are only at below 50% of their initial survey level. Between 2005 and 2014, shrimp farms have declined by approximately 400 ha with some loss in the south and what appears to be some ribbon mangrove forest extensions throughout. Again, these ribbon forests were likely present throughout the survey period, but the improved resolution of current satellite imagery systems is now able to capture these narrow but continuous mangrove forests adequately. The other LULC class in Muisne Estuary remained stable between 2005 and 2014. As of 2014, Muisne Estuary is a shrimp farm dominated estuary, but mangrove forests have made a small recovery from their low levels of the 1990s.

Figure 5.4 depicts the percentage of each LULC in Muisne Estuary. As of 1970, mangrove forest LULC was 52% of the estuary's LULC and reached 24% in 2014 (Fig. 5.4). The majority of mangrove forest loss occurred between 1986 and 1998, but since 1998, mangrove forest has been expanding, albeit modestly. Shrimp farms began with no presence in the estuary by study design and the period ended with shrimp farms consuming 36% of the estuary (Fig. 5.4). Almost all the shrimp farm expansion occurred between 1986 and 1998, and since then, shrimp farms levels

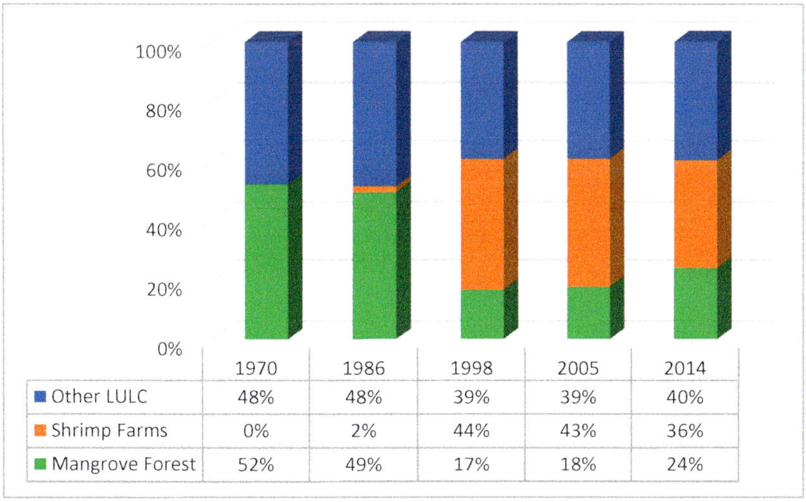

Fig. 5.4 Muisne estuary LULC change over time

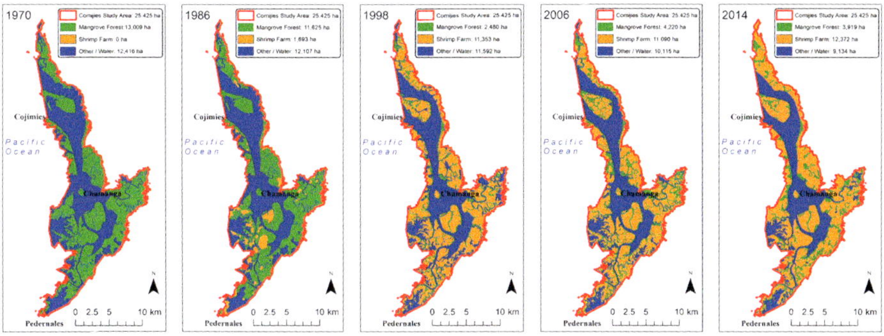

Fig. 5.5 Cojimies estuary LULC change from the arrival of shrimp aquaculture to 2014

have remained constant and even began to decline into the 2010s. The other LULC class lost area between 1986 and 1998, but since then has remained at a steady level through 2014. In addition to the survey years presented in this chapter, a survey from 2011 is available in the Dataverse repository.

5.1.3 *Cojimies Estuary*

Figure 5.5 depicts the three LULC classifications throughout Cojimies Estuary from the period before the arrival of shrimp farms in 1970 through 2014. In 1970, Cojimies Estuary had no shrimp farms present and was dominated by mangrove forests and the other LULC class. Mangrove forests covered 13,009 ha of the estuary with substantial mangrove forests covering the interior of the estuary (Fig. 5.5). The other LULC designation, comprised mostly of surface water, salt-pans, and mudflats, occupied 12,416 ha of estuarine space. In 1970, Cojimies Estuary had both substantial and contiguous mangrove forests with the interior portions of the estuary and some more limited in extent, but still relatively large mangrove stands, near the mouth of the estuary.

By 1986, shrimp farms had appeared within Cojimies Estuary, but it is still dominated by mangrove forests and the other LULC designation (Fig. 5.5). The shrimp farms appeared in the south, mostly located on isolated islands in the deep interior of the estuary. The 1986 shrimp farm coverage in the estuary is only 1693 ha out of a study area size of 25,425 ha. The shrimp farms that did appear by this time reside mostly on islands in the interior portion of the estuary (Fig. 5.5). Mangrove forest remains the dominant LULC covering 11,625 ha of the estuary, and the other LULC classification remains approximately stable at 12,107 ha (Fig. 5.5). Cojimies remained a mangrove forest dominated estuary in 1986, but shrimp farms become numerous on the islands in the southern portion of the estuary.

Between 1986 and 1998 the mangrove forests of Cojimies Estuary appear to have undergone substantial clearing. In 1986, 11,625 ha of mangrove forests existed in the estuary, but by 1998, have declined to only 2480 ha (Fig. 5.5). Cojimies Estuary lost almost 80% of its once intact mangrove forests in these 12 years. Conversely, by 1998, shrimp farms have expanded to a high level in Cojimies Estuary (Fig. 5.5). Shrimp farms consumed only 1693 ha of estuary space in 1986, but a little over one decade later, they cover 2842 ha of the estuary (Fig. 5.5). Shrimp farms are the dominant LULC of the estuary at this time, consuming more estuary space than either mangrove forest or the other LULC classification. As with neighboring Muisne Estuary, it would now be correct to state that some limited mangrove forests exist in a shrimp farm environment, as opposed to the more traditional designation of shrimp farming occurring in the mangrove forest environment. The other LULC class appears to have undergone some nominal decline between 1986 and 1998 losing 515 ha (Fig. 5.5).

As with neighboring Muisne Estuary, between 1998 and 2006, the shrimp farm area of Cojimies Estuary remains inherently stable (Fig. 5.5). There does appear to be a trend of mangrove forest gain in the eastern interior portion of Cojimies Estuary between these dates. The new forests appear to snake around the periphery of preexisting shrimp farms (Fig. 5.5). Four possible explanations exist for this new mangrove forest in Cojimies. Firstly, it is a new mangrove forest. Secondly, the pattern indicates it may be long and narrow forests between the shrimp farms and the river. It is possible that improved instrument resolution could have detected this forest and it was there all along. The third possibility is that it is a terrestrial forest that has crossed into the estuary and been misclassified, and this appears likely as the next survey removes portions of this forest. Cojimies is a heavily degraded estuary with sedimentation and water circulation issues, and the estuary is dry by 2006 in many interior portions. Finally, the image may be at extremely low-tide, or the previous survey was at high tide. If this occurred and substantially less water exists in and around the mangrove forest edge, then the mangrove forest class is likely to have increased reflectance in the infrared bands and appear to increase in size. This is one of the few areas of potential mangrove forest expansion that was not field verified, so the assumption is that it is new mangrove forest growth until shown otherwise. The new mangrove forest area in 2006 in Cojimies Estuary is 4220 ha, and although relatively small in real terms, it is a substantial increase as a percentage of 1998 mangrove forest area in this estuary. Despite this mangrove forest increase, shrimp farms remain the singular dominant LULC in Cojimies Estuary in 2006 (Fig. 5.5). This region additionally has other LULC class losses during this period.

In the period between 2006 and 2014, shrimp farms in Cojimies Estuary have increased by approximately 1300 ha, mangrove forests have declined somewhat from 2006, losing approximately 300 ha of cover, and the other LULC class has diminished by approximately 1000 ha (Fig. 5.5). The final shrimp farm area in Cojimies Estuary is 12,372 ha which is more than either mangrove forest or the other LULC class and almost as much as both combined. The final mangrove forest level in Cojimies Estuary is 3919 ha which is approximately 30% of its initial survey level (Fig. 5.5). By 2014, Cojimies Estuary is a shrimp farm dominated environment with only isolated pockets of mangrove forest remaining.

Figure 5.6 depicts the percentage of each LULC in Cojimies Estuary by year. Before the introduction of aquaculture, mangrove forest LULC was 51% of the LULC in the estuary in 1970, but by 1998, had dropped to only 10%. Mangrove forests recovered somewhat to 15% of estuarine LULC by the end of the survey period, but mangrove forests remain at only a small fraction of their pre-aquaculture area. Most of the mangrove forest loss occurred before 1998, and from 1998 to 2014, the mangrove forest LULC has been on a slight upward trend, but this is starting from an exceptionally low level.

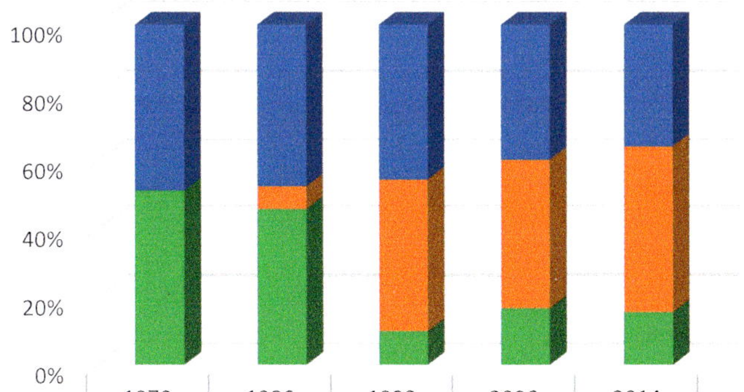

	1970	1986	1998	2006	2014
Other LULC	49%	48%	46%	40%	36%
Shrimp Farms	0%	7%	45%	44%	49%
Mangrove Forest	51%	46%	10%	17%	15%

Fig. 5.6 Cojimies estuary LULC change over time

Conversely, shrimp farms were not present in the estuary in 1970 but were the dominant LULC by 2014, and they now consume almost half of all estuarine space (Fig. 5.6). From gaining a small foothold in the estuary in 1986, shrimp farms became the dominant LULC within 12 years. Most of the shrimp farms arrived in the estuary between 1986 and 1998, and since 1998, shrimp farms have continued to consume estuarine space but at a much-reduced rate (Fig. 5.6). The other LULC class in Cojimies, which is almost all surface water in this study site, has had moderate declines in area coverage across the entire study period. In addition to the survey years presented in this chapter, an additional survey from 2011 is available in the Dataverse repository.

5.1.4 Chone Estuary

Figure 5.7 depicts the three LULC classes throughout Chone Estuary from the period before the arrival of shrimp farms in 1968 through 2014. Chone Estuary is oversampled to give more temporally explicit information between the decadal periods. In 1968, Chone Estuary had no shrimp farms present and was dominated by mangrove forests and the other LULC designation, which in this estuary is comprised of water and the terrestrial environment. In 1970, the estuary was comprised of 4207 ha of mangrove forest, 4448 ha of other LULC, and no shrimp farms had yet been constructed (Fig. 5.7). The mangrove forest is most prominent in the eastern interior of the estuary, but some large stands are located near the mouth of the estuary. Mangrove forest is the dominant LULC, and the other LULC designation is also substantial.

By 1977 (Fig. 5.7) three large shrimp farm clusters appeared in the eastern interior of Chone Estuary. The size of the ponds indicates that these pioneer ponds are likely managed in an extensive manner. The 1977 datapoint in Chone is one of the oversampled points to give insight between the decadal periods and because it was known that only a few large and extensive ponds appear in the 1970s, so the mass expansion of semi-intensive aquaculture must have occurred after this period. In 1977, the nascent shrimp farm industry consumed only 324 ha of the estuary. At this time, Chone Estuary was still a mangrove forest dominated estuary with approximately 3843 ha of mangrove forest present. The other LULC designation remained consistent from 1968 at 4448 ha (Fig. 5.7).

Between 1977 and 1984 shrimp farm area exploded in Chone Estuary (Fig. 5.7). In 1977, the nascent shrimp farm industry consumed only 324 ha of a small portion of the interior of the estuary, but by 1984 the shrimp farm area has grown to 3642 ha in area, and growout ponds are present across almost the entire estuary. Shrimp farm area has expanded over 11-fold during this period, from being just a cluster of ponds in 1977 to consuming more estuary space then either mangrove forest or the other LULC class (Fig. 5.7). During the period between 1977 and 1984, mangrove forest appears to have undergone substantial clearing. In 1977, mangrove forests dominated the estuary covering 3843 ha. By 1984, the mangrove forest had decreased by

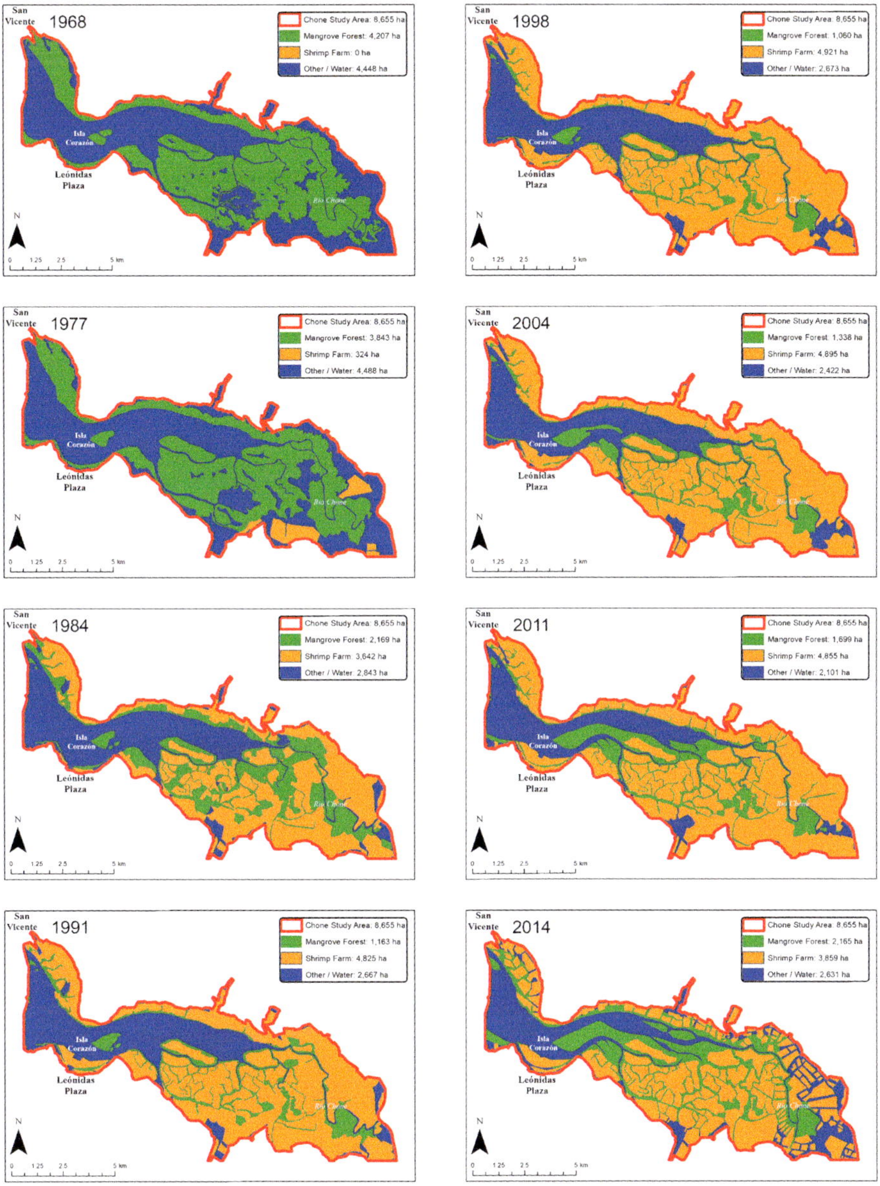

Fig. 5.7 Chone estuary LULC change from the arrival of shrimp aquaculture to 2014

44% to 2169 ha. This mangrove forest clearing occurred in only 7 years. Not only mangrove forests diminished in the estuary between 1977 and 1984. During this period, the other LULC class diminished from 4488 ha to 2843 ha, constituting a 37% loss (Fig. 5.7).

Between 1977 and 1984 Chone Estuary transitioned from a mangrove forest dominated estuary to a shrimp farm dominated estuary. In addition to this transition, the estuary itself appears to be shrinking as the other LULC class, which is greater than 90% water with some areas of salt pans and mudflats, has decreased substantially in size. By 1984, shrimp farms not only occupied more estuarine space than either mangrove forest or the other LULC class but they also now consume more estuarine space than mangrove forest and all other LULC classes combined, including surface water (Fig. 5.7). In 7 years, shrimp farms went from a nascent industry in the region to occupy most of the estuarine real estate.

Between 1984 and 1991, the LULC change pattern in Chone Estuary is a continuation of the period between 1997 and 1984. During the 1984–1991 period, shrimp farm area continues to expand, mangrove forests continue to decline, and the other LULC continues to decline at a reduced rate (Fig. 5.7). By 1991, mangrove forest area has decreased from the 1984 level by another approximately 1000 ha, and mangrove forest coverage diminished by 73% of its initial 1969 area. Shrimp farms continue to expand from 3642 ha in 1984 to 4825 ha in 1991 (Fig. 5.7). The pattern between 1984 and 1991 is of shrimp farms filling in around ponds and expanding farther out into the estuary. Some additional other LULC loss continued to occur but at a much-reduced level than the earlier period.

The 1990s in Chone Estuary appears to be a period of relative stability after the earlier periods of rapid LULC change. The estuary has entered a new normal state of shrimp farms dominating all other LULC classes including mangrove forest. Mangrove forests lost an additional 10% of their area, down in coverage area from 1163 ha in 1991 to 1060 in 1998 (Fig. 5.7). This reduced mangrove forest loss rate appears to be partially due to only two significant areas of forest remaining to be deforested. Shrimp farm LULC remains inherently stable during this period. Shrimp farms still occupy more estuarine space than all other LULC types combined including mangrove forests.

Between 1998 and 2004, shrimp farm coverage remains stable and the estuary remains dominated by shrimp farms covering more estuarine space than all other LULC types combined. However, a new LULC phenomenon appears in the 2004 estuary survey, the beginnings of mangrove forest recover (Fig. 5.7). Mangrove forest appears to be expanding in and around Isla Corazon near the mouth of the estuary (Fig. 5.7). This new mangrove forest extension drives an increase in estuarine mangrove forest cover, with the 2004 level of mangrove forest increasing to 1338 ha from the 1998 level of 1060 ha. This regrowth was field-verified in 2007 as a *Rhizophora mangle* replanting conducted mainly by an artisanal fishing community in the north of the estuary. During the 1998–2004 period, the other LULC class dropped by about 200 ha. The new mangrove forest growth needs to be put into perspective; mangrove forest cover in 2004 in the Chone Estuary remains at less than one-third of the initial survey level (Fig. 5.7).

Between 2004 and 2011, the trend of LULC change in Chone Estuary continues the trajectory of the previous period. Shrimp farming in the estuary remains stable, but mangrove forests continue to gain more coverage (Fig. 5.7). The extension of mangrove forest that started around Isla Corazon in the earlier period has continued much further east into the interior of the estuary. With this data point, mangrove forest now has two consistent periods of recovery documented in Chone Estuary. This mangrove forest regrowth was field-verified in 2012 as the same *Rhizophora mangle* replanting noted in the earlier period. Mangrove forest has now increased to 1699 ha in 2011, from a low of 1060 ha in 1998 (Fig. 5.7). This is a 70% increase in mangrove forest during the 13 years. Mangrove forest has now recovered to 40% of its pre-shrimp farm level measured in 1968 (Fig. 5.7). During this period of mangrove forest regrowth, the other LULC class, consisting mostly of water, continues its steady decline from 2422 ha in 2004 to 2101 ha in 2011. Indeed, the other LULC class is 40% of its pre-shrimp farm initial survey level; this is the same total rate of loss that mangrove forest suffered from the initial survey to 2011. Most of the other LULC class loss is in the western interior of the estuary and on and around Isla Corazon (Fig. 5.7). The mangrove forest recovery, although now significant, must be put into context. The shrimp farm LULC class still covers more Chone Estuary area in 2011 than all other LULC classes combined. The estuary is still a majority shrimp farm environment despite the mangrove forest regrowth.

The transition to VHR satellite data in 2014 (Table 4.1) reveals some interesting facets to the LULC change analysis of Chone Estuary. In 2014, shrimp farms appeared to undergo a substantial decline in the interior eastern portion of the estuary (Fig. 5.7). In this same region, new long and narrow linear mangrove forests have appeared. This new mangrove forest was field-verified in 2015, and it does exist but is likely not new growth. The higher resolution spatial data reveals a long continuous band of mangrove trees, constituting a substantial area when combined, that exist on the edge of the shrimp farms between them and the river or estuarine waters. The ribbon mangrove stands that encircle the ponds were too small in width to be detected by the lower resolution earlier instruments but are now captured by the higher-resolution Rapid Eye instrument although they likely always existed but were just not recorded in earlier surveys. These 'new mangroves' now contribute to the mangrove forest total of 2577 ha in 2014. The mangrove forest expansion in 2014 is not all instrument-derived, however. New mangrove stands have appeared to the northeast of Isla Corazon as well (Fig. 5.7), and this new growth was field-verified in 2015. Despite this mangrove forest expansion, most of the new mangrove forest in the interior appears to be due to improved instrument resolution. These newly detected ribbon mangrove forests were field-verified in 2016 and are present. This VHR-derived mangrove forest in Chone Estuary remains comparable to the initial survey as the regions originally had this area of mangrove forest detected in 1968 before the shrimp farms arrived. The final mangrove forest area calculation for Chone Estuary in 2014 is 2165 ha, and this constitutes a 100% recovery from the mangrove forest low period of 1998 when only 1060 ha of mangrove forest remained (Fig. 5.7). Mangrove forests have now recovered to 61% of their initial survey level.

Shrimp farm coverage in the 2014 Chone Estuary survey is 3859 ha (Fig. 5.7). This is an approximate 1000 ha reduction since 2011. Two or three shrimp farms in the northern portion of the estuary have returned to mangrove forest LULC, but most of this shrimp farm decline is due to the instrument resolution change described above. Mangrove forests along the edge of the shrimp farms can now be delineated,

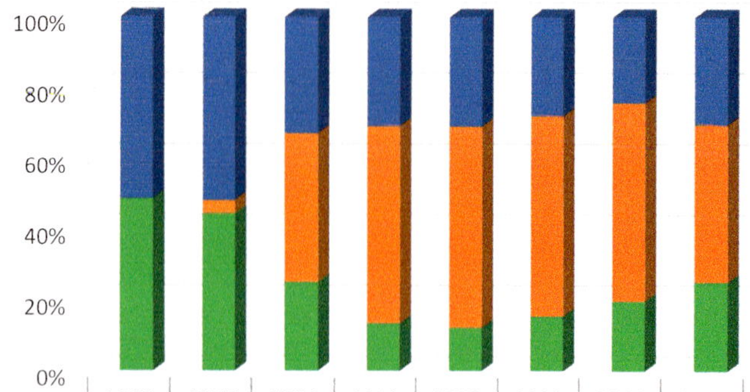

	1968	1977	1984	1991	1998	2004	2011	2014
Other LULC	51%	52%	33%	31%	31%	28%	24%	30%
Shrimp Farms	0%	4%	42%	56%	57%	57%	56%	45%
Mangrove Forest	49%	44%	25%	13%	12%	15%	20%	25%

Fig. 5.8 Chone estuary LULC change over time

and this has resulted in a decrease in shrimp farm area between 2011 and 2014. The other LULC class has additionally benefited from the change in the instrument, as river channels around the shrimp farms in the northwest corner of the estuary are now detectable, resulting in an increase in the water area within Chone Estuary. Again this change is likely 100% instrument-driven but does capture a more nuanced LULC in the estuary that is likely closer to the ground reality.

Figure 5.8 shows the LULC of Chone Estuary for each survey period. Mangrove forests shrunk in size from the initial survey through 1998, but from 1998 onwards they have recovered somewhat and show a long-term and consistent pattern of expansion in the estuary. Shrimp farms first appeared and expanded to about 1991. From 1991 to 2011 they essentially remained stable until reducing in size in 2014 (Fig. 5.8). The reduction in shrimp farm area from 2011 to 2014 is primarily instrument related, but the percentages provided by the VHR instrument are likely representative of the actual LULC present in the Chone Estuary. The other LULC class, consisting of 99% water, has been in almost constant decline since the initial survey. That is, the estuary has less area covered by surface water, salt-pans, and mud flats. This reduction continued year to year until the 2001–2014 period which saw a slight increase in the other LULC class that was again mostly instrument driven change. This limited increase in other LULC during the final survey period is likely due to the improvement in the spatial resolution of the VHR instrument than actual recovery.

Despite substantial real and field-verified mangrove forest recovery and the instrument-driven decline in shrimp farms and an increase in the other LULC class, Chone Estuary remains a shrimp farm environment as of 2014. Shrimp farms still occupy more estuary real estate than either mangrove forests or the other LULC class. In addition to the eight surveys discussed in this book, two additional surveys are provided on the accompanying Dataverse website. These two additional surveys are for 2008 and 2011 respectively.

5.1.5 *Isla Puná North*

In 1977, the LULC of Isla Puná North consisted of 1501 ha of mangrove forest, 0 ha of shrimp farm landcover, and 5653 ha of the other LULC class which in this area is mostly water, other wetlands, salt-pans, and mudflats (Fig. 5.9). Most of the mangrove forest exists in a thin ribbon-like strip in a lagoon-like environment off the north coast of Isla Puná. Some other mangrove forests exist along river mouths on the landward edge of the lagoon-like environment. By1985, three large clusters of shrimp ponds arrived in the study area: one small cluster in the west of the study area, another small cluster towards the east of the study area, and one large cluster of farms in the center of the study area. The total area of the newly arrived shrimp farms is 606 ha (Fig. 5.9). Between 1977 and 1985, mangrove forest LULC decreased by approximately 400 ha from 1501 ha to 1098 ha and the other LULC class remained consistent during this period.

Between 1985 and 1990 in the Isla Puná North study area, mangrove forests remained mostly stable, gaining approximately 100 ha, while the other LULC class lost 627 ha of its LULC (Fig. 5.9). During these 5 years, shrimp farms almost doubled in area from 606 ha in 1985 to 1136 ha in 1990. Most of this increase appears to be an expansion and filling-in of the landscape between the ponds present in 1985 and some more modest expansion on the eastern and western ends of the study area (Fig. 5.9). Shrimp farms LULC has just exceeded mangrove forest LULC in the estuary by 1990.

Between 1990 and 2000, mangrove forest in the study area remains inherently stable, decreasing slightly from 1195 ha to 1116 ha, whereas shrimp farms expand a little under 300 ha from 1136 ha in 1990 to 1413 ha in 2000 (Fig. 5.9). Once again, the other LULC class loses some area, decreasing by approximately 200 ha. Between 2000 and 2008, mangrove forest extends to 1314 ha, which is less than 200 ha from its initial survey level (Fig. 5.9). Shrimp farms continue their gradual expansion increasing in area from 1413 ha in 2000 to 1716 ha in 2008. The other LULC class continues its decline with somewhat of an acceleration, losing 500 ha of LULC within the Isla Puná North study area.

Between 2008 and 2014, mangrove forests again remained mostly the same with a slight increase in area of just under 100 ha (Fig. 5.9). The small gains noted over numerous surveys have resulted in mangrove forest now occupying almost as much space as they did in the initial 1977 survey. Between 2008 and 2014, shrimp farms continue to increase from 1716 ha to 2015 ha. The gradual increase in shrimp farms throughout the study period in the Isla Puná North study area has resulted in the construction of 2104 ha of shrimp farms. The other LULC class continues the pattern of steady decreases between 2008 and 2014, losing approximately 400 ha within the study area (Fig. 5.9). This has resulted in the other LULC class losing

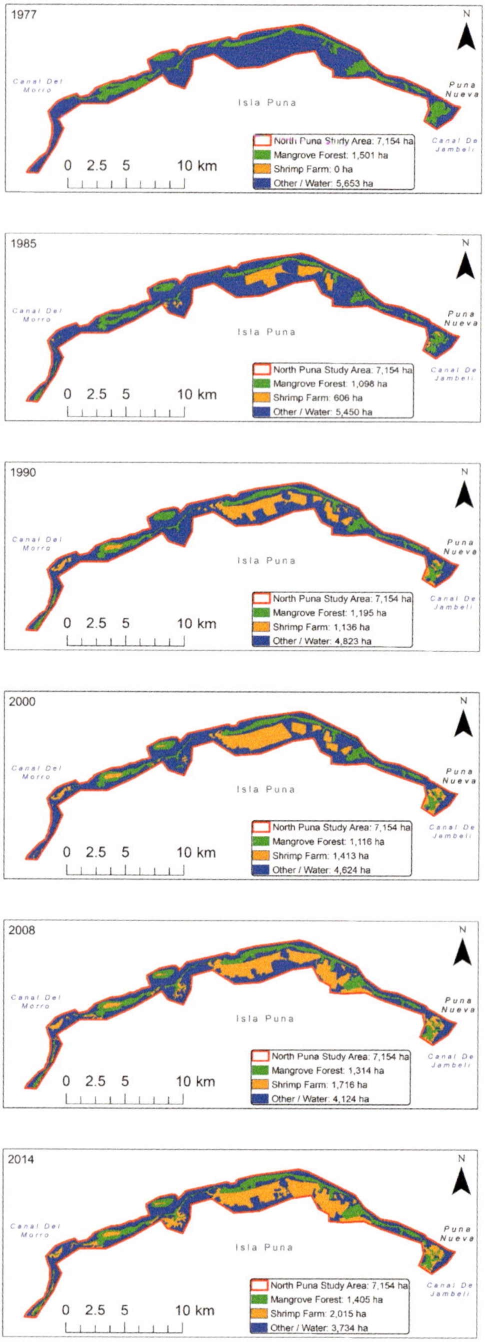

Fig. 5.9 Isla Puná north estuary LULC change from the arrival of shrimp aquaculture to 2014

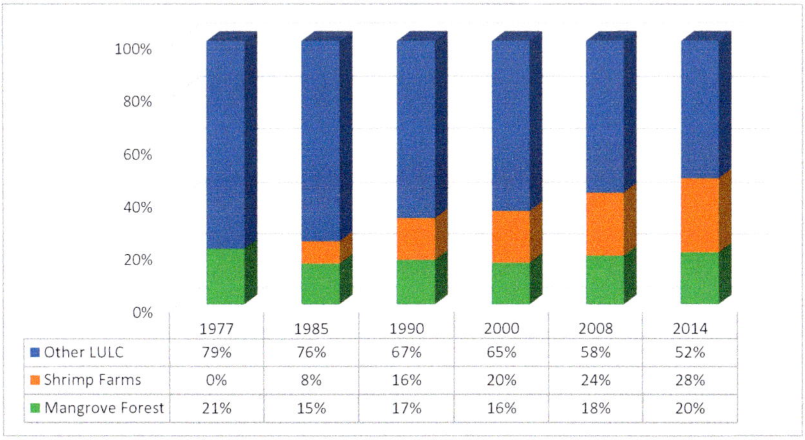

Fig. 5.10 Isla Puná north estuary LULC change over time

almost 2000 ha of space across the entire analysis period. As of 2014, the other LULC class is still the majority landcover in the estuary, but shrimp farms are starting to approach this level while the mangrove forest remains mostly intact.

Figure 5.10 depicts the relative proportions of each LULC that constitute the Isla Puná North study area across the entire analysis period. Mangrove forest suffered some limited early losses in area but gradually recovered throughout the analysis period, nearly reaching their initial level by the end of the analysis period. Within the Isla Puná North study area, shrimp farms expanded at a consistent rate from decade to decade and by the end of the analysis period consume substantially more of the study area than mangrove forests (Fig. 5.10). The other LULC class consistently decreased across the Isla Puná North study area from decade to decade. In 1970, the other LULC class occupied 79% of the study area, but by 2014, it decreased to just over half (Fig. 5.9).

5.1.6 *Isla Puná*

In 1973, pre-shrimp farm arrival LULC of Isla Puná was constituted of 9947 ha of mangrove forest, 0 ha of shrimp farms, and 21,297 ha of the other LULC class which in this area is mostly water, salt-pans, mudflats, and some limited terrestrial land (Fig. 5.12). Most of the mangrove forest exists in a single and large estuary along the western coastline of Isla Puná named Estero Barbascal located on the Jambelí Channel facing the Ecuadorian mainland. Outside of the large estuary, mangrove forest exists in another small estuary to the south of the main one named Estero de la Aguada and in a ribbon-like manner extending away from the main estuary to the north at the mouth of the Rio Hondo (Fig. 5.12).

By1985, three clusters of shrimp ponds arrived in the estuary: one small cluster in the north of estuary near the mouth of the Rio Hondo, one large cluster of shrimp farms in and around Estero Barbascal, and one large cluster of shrimp farms in Estero de la Aguada. The total area of the newly arrived shrimp farms is 2792 ha (Fig. 5.12). Between 1977 and 1985, mangrove forest LULC remained entirely stable while the other LULC class decreased in size from 21,297 ha to 18,506 ha (Fig. 5.12).

Between 1985 and 1997 in the Isla Puná study area, mangrove forests diminished in size by 618 ha (Fig. 5.12). During this same period, the other LULC class underwent a significant decrease in size, losing 6500 ha of its LULC (Fig. 5.12). Much of this loss appears to have occurred in and around the wetlands and salt-pans of Estero de la Aguada. During these 12 years, the other LULC class dropped by a little over 35%. During this same period, shrimp farms consumed almost all the estuarine space in and around Estero de la Aguada, increasing in area from 2792 ha in 1985 to 9910 ha in 1997; this equates to a 350% increase in size in these 12 years. Very little of the southern portion of the estuary area remains without a shrimp farm (Fig. 5.12).

Between 1997 and 2008, mangrove forests continue their pattern of a slow but steady decline in the Isla Puná study area. During this period, mangrove forests decrease in size from 9329 ha to 8793 ha, which is a loss of 536 ha (Fig. 5.12). The other LULC class also continues its decline from the earlier period, losing another 1672 ha of estuarine space between 1997 and 2008. During this period, shrimp farms continue their expansion, increasing their area by another 2208 ha. By 2008, shrimp farms are now the majority LULC across the study site, surpassing both mangrove forest and the other LULC class. By this time, the study area is a shrimp farm dominated environment with some intermittent other landcover including mangrove forest present.

Between 1997 and 2008, the prior period's trend of mangrove forest loss, shrimp farm expansion, and other LULC class loss continued, though smaller in magnitude. Mangrove forests declined by 640 ha, shrimp farms expanded by 1085 ha, and the other LULC class declined in area by 445 ha (Fig. 5.12). The Isla Puná study area remains mangrove forest dominated in 2014 with mangrove forests almost exclusively confined to the interior eastern portion of Estero Barbascal (Figs. 5.11 and 5.12).

The Isla Puná study area has undergone remarkable consistent LULC changes across all time frames from the initial survey in 1973 to the final survey in 2014. Mangrove forest LULC has consistently declined from 1973 to 2014, albeit at a relatively slow rate (Fig. 5.12). In the initial survey, mangrove forest occupied 32% of estuarine space, but by 2014, had decreased to 26% of estuarine space. Shrimp farms have consistently expanded in area across all periods with a substantial increase between 1985 and 1997, followed by more moderate increases. In the initial survey, shrimp farms occupied no estuarine space but by 2014, had increased to 42% of estuarine space, making them the dominant LULC in the study area, covering more area than mangrove forests or the other LULC class. The other LULC class has had consistent decade over decade declines from 1973 to 2014 which

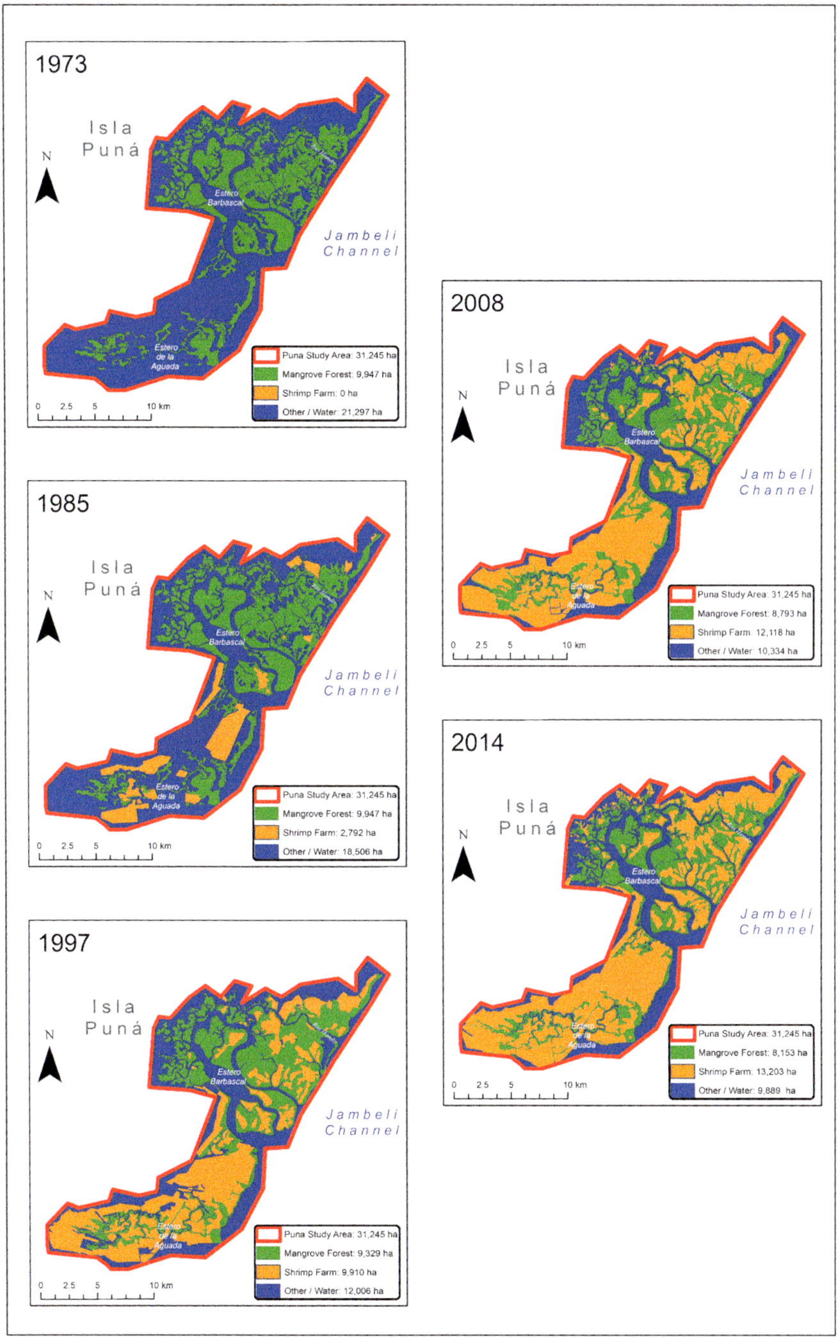

Fig. 5.11 Isla Puná estuary LULC change from the arrival of shrimp aquaculture to 2014

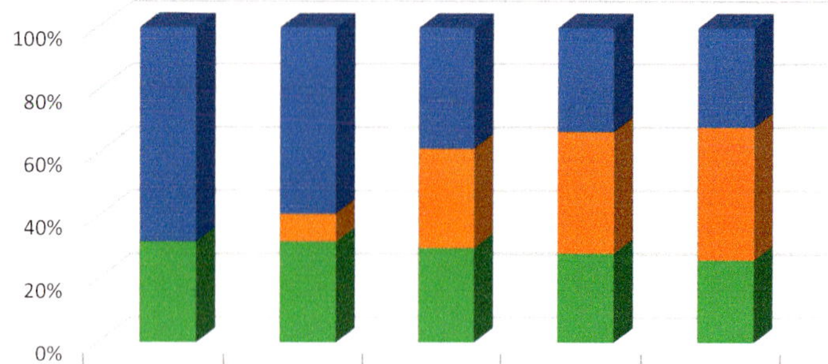

	1973	1985	1997	2008	2014
Other LULC	68%	59%	38%	33%	32%
Shrimp Farms	0%	9%	32%	39%	42%
Mangrove Forest	32%	32%	30%	28%	26%

Fig. 5.12 Isla Puná estuary LULC change over time

indicates a decrease in the amount of estuarine space (Fig. 5.12). At the initial survey, the other LULC class occupied 68% of estuarine space, but by 2014, this had decreased by more than half to occupy only 32% of estuarine space. Additional surveys for 1990 and 2011 are included in the accompanying Dataverse website.

5.1.7 Guayas Estuary

As of 2019, the pre-1980s aerial photography and 1:25,000 topographic remain incomplete for Guayas Estuary, so no full pre-aquaculture survey is possible. As opposed to all other study areas, substantial shrimp farming has already arrived by the time of the initial survey in 1985. At that time, Guayas Estuary consisted of 134,035 ha of mangrove forest, 30,800 ha of shrimp farms, and 125,142 ha of the other LULC class that is mostly water and some limited terrestrial environment (Fig. 5.13). When compared to the northern estuaries, all the of the LULC classes are substantially larger. Even with this amount of shrimp farming present, the estuary is still dominated by mangrove forests, and the western edge and center of the estuary are heavily forested.

Between 1985 and 1990, mangrove forests in Guayas Estuary decreased in extent by just under 10,000 ha and now cover 124,351 ha of the estuary (Fig. 5.13). The mangrove forest decrease occurred across all portions of the estuary. During this same period, shrimp farms increased in the Guayas Estuary from 30,800 ha in 1985 to 47,881 ha in 1990. This is more than a 50% increase in shrimp farm area over 5 years. The other LULC class decreased by 7397 ha from 125,142 ha in 1985 to 117,745 ha in 1990 (Fig. 5.13).

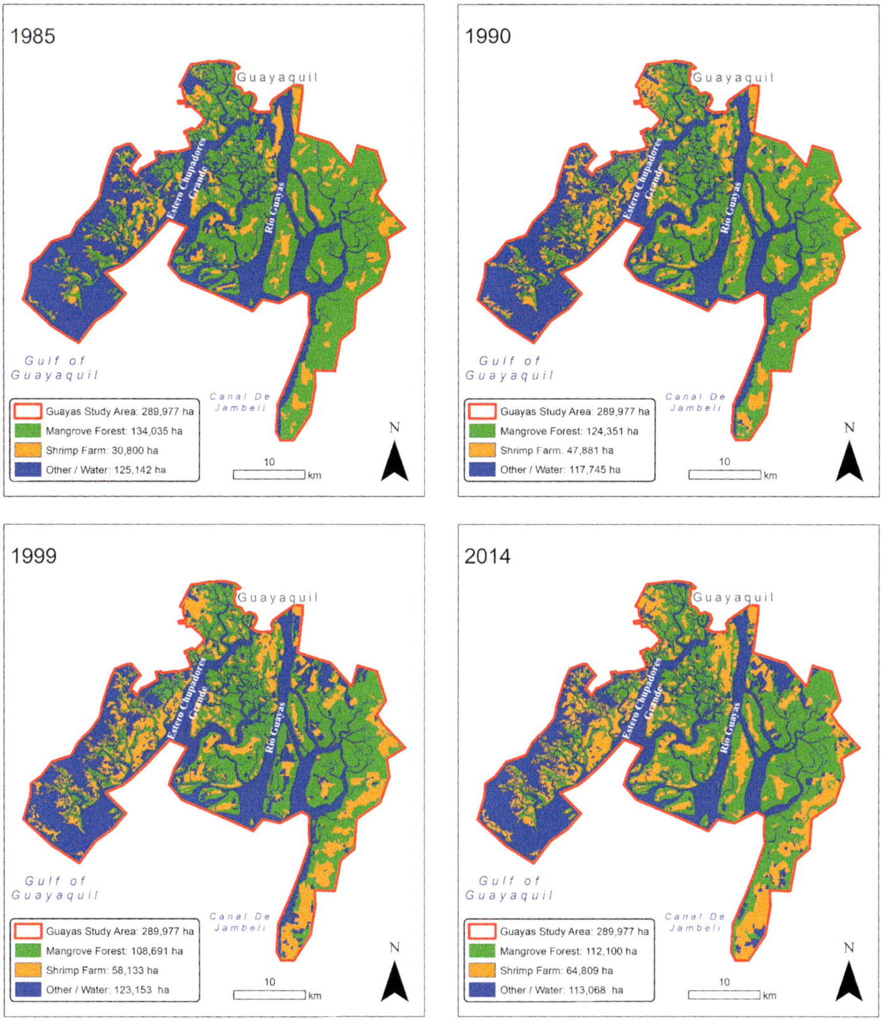

Fig. 5.13 Guayas estuary LULC change from the arrival of shrimp aquaculture to 2014

During the decade between 1990 and 1999, the trends of mangrove forest area decreasing and shrimp farm extent increasing continued. In 1990, mangrove forest lost an additional 15,660 ha of coverage declining from 124,351 ha in 1995 to 108,691 in 1999 (Fig. 5.13). During this same period, shrimp farms expanded from 47,881 ha in 1990 to 58,133 ha by 1999; this is an additional 10,252 ha. The other LULC class covered 123,153 ha of estuary space in 1999 and is less than 2000 ha, or within 1.6% of its initial 1985 area (Fig. 5.13).

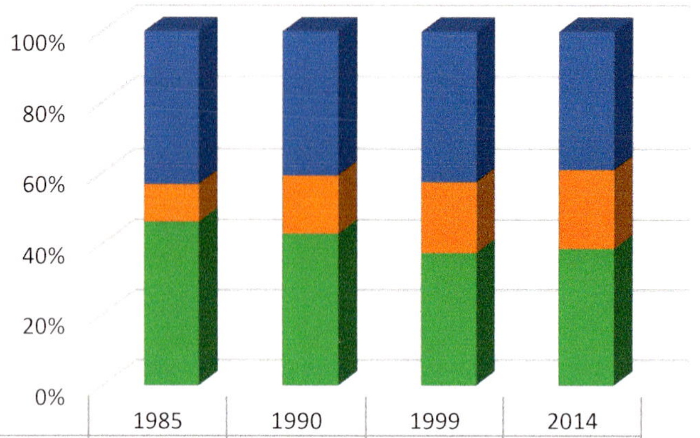

	1985	1990	1999	2014
Other LULC	43%	41%	42%	39%
Shrimp Farms	11%	17%	20%	22%
Mangrove Forest	46%	43%	37%	39%

Fig. 5.14 Guayas estuary LULC change over time

During the 15 years between 1999 and 2014, the long-term trend of decreasing mangrove forest extent appears to have ended, but the long-term trend of increasing shrimp farm extent continues. By 1999, the mangrove forest area of the estuary was 112,100 ha, and this is 3409 ha more mangrove forest area than what existed in 1999 (Fig. 5.13). Mangrove forests have declined from the initial survey in 1985 to the final survey in 2014 by 21,935 ha or 16%. During the period between 1999 and 2014, shrimp farms continue to increase in area but by a much-reduced rate when compared to earlier periods. By 2004, shrimp farms occupied 64,809 ha of estuarine space from their initial and already high level of 30,800 ha (Fig. 5.13). Shrimp farms more than doubled in extent in Guayas Estuary between 1985 and 2014. In this final analysis period, the other LULC class has lost over 10,000 ha since 1999 (Fig. 5.13). By 2014, Guayas Estuary remains a mangrove forest and other LULC class dominated estuary, but the shrimp farm area is now far greater than 50% of the mangrove forest area.

Within Guayas Estuary, the three LULC classes have a somewhat consistent change trajectory over the analysis period. Mangrove forests decline from covering 46% of the estuary in 1985 to 39% in 2014 (Fig. 5.14). Despite this overall decrease, the final analysis period points to mangrove forest recovery in Guayas Estuary, but another survey is required post-2014 to ascertain if this increase is an anomaly or a true longer-term reforesting of mangroves. Across the entire analysis period, and at each observation point, shrimp farms continued to increase in area across Guayas Estuary (Fig. 5.14). The rate of shrimp farm increase decreases over time. By 2014, shrimp farms occupy 22% of Guayas estuary from an initial level of 11%. The other LULC class shows a moderate decrease across the analysis period, from covering 46% of estuarine space in 1985 to covering 39% in 2014.

5.1.8 The Estuaries of El Oro Province

The initial survey for El Oro province uses a suite of topographic maps from 1977 and prior, creating a pre-shrimp farm initial survey of the estuaries of El Oro Province. As with Guayas Estuary, some shrimp farms are visible in the earliest survey, but these nascent shrimp farms constitute only 5% of the entire study area. The actual study area of El Oro crosses into Guayas Province at its northern end to avoid splitting a singular estuary in two. The two study areas of Guayas Province and the Estuaries of El Oro Province meet at a shared border and do not overlap.

In 1977, the Estuaries of El Oro Province contained 32,154 ha of mangrove forest, 3451 ha of shrimp farms, and 33,519 ha of the other LULC class (Fig. 5.15). The other LULC class consists of a wide variety of differing landcovers long the El Oro coastline including wetlands, salt-pans, mudflats, and some terrestrial environment encompassing parts of the urban environment of Machala. The southern portion of the study area consists mainly of mangrove forests in and around Grande Estuary which extends some distance inland from the Gulf of Guayaquil. North of Machala, the topography is less low-lying with more relief, and thus the potential mangrove forest area changes to a narrow strip of coastline.

By 1985, the mangrove forests of El Oro Province have undergone substantial declines in their extent. The mangrove forest area decreased by a little over 8000 ha, or 25% of their original extent (Fig. 5.15). The mangrove forests north of Machala to the Guayas border are almost all gone in the 1985 survey, and many of the mangrove forests in the southeast of the study area are either much reduced in size or cleared. Conversely, shrimp farms underwent dramatic increases in extent during the 1977–1985 period. Shrimp farms cover 25,685 ha of study area space in 1985, and their footprint across the study site is larger than that of the mangrove forest (Fig. 5.15). Shrimp farms increased over seven-fold during these 8 years. The other LULC class underwent more substantial losses than mangrove forests, from 33,519 ha in 1977 to 19,351 ha in 1985 (Fig. 5.5). The other LULC class underwent a 58% reduction during these 8 years. By 1985, shrimp farms occupy more space than either mangrove forests or the other LULC class.

By 1997 the transition to a shrimp farm environment is complete across the estuaries of El Oro Province. Shrimp farms expanded between 1985 and 1997 by just under 10,000 ha and now occupy more of the study area than mangrove forests and all other LULC classes combined. In 1997, shrimp farms covered 35,300 ha of the 69,124 ha study area (Fig. 5.15). Between 1985 and 1997, mangrove forests continue their downward trajectory. In 1985, they covered 24,088 ha of the study area but decreased to 17,885 ha by 1997. Mangrove forests occupy a little over half the area they covered in the initial survey. The other LULC class continued its downward trajectory between 1985 and 1997. It covers 15,939 ha in 1997, down from 19,351 ha in 1985 and 33,519 ha in 1977 (Fig. 5.15).

The period between 1997 and 2006 is one of general LULC stability after 20 years of rapid and substantial change. The Estuaries of El Oro Province appear to have entered a new normal phase of shrimp farm dominance with only slight

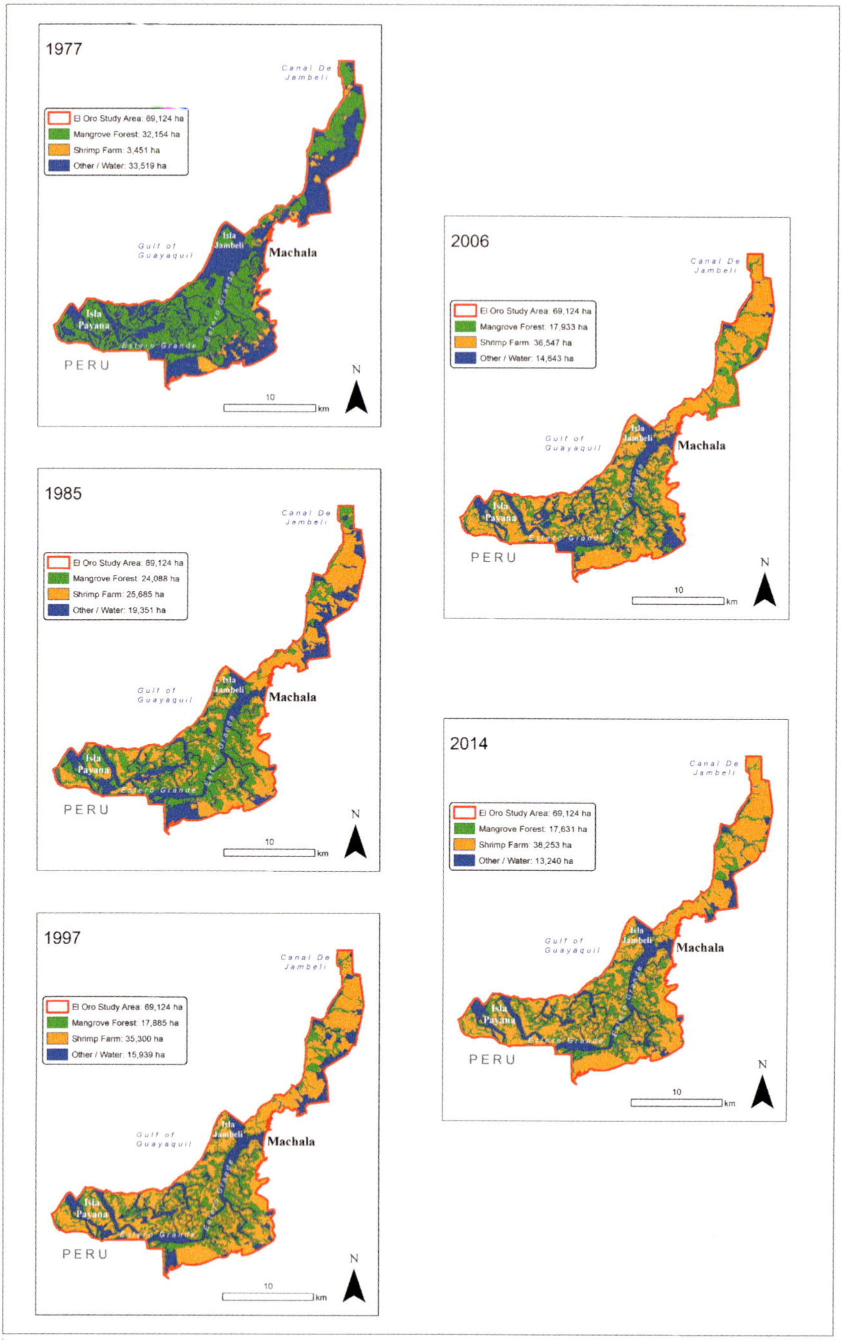

Fig. 5.15 The estuaries of El Oro province LULC change from the arrival of shrimp aquaculture to 2014

changes in LULC occurring. Mangrove forests remain stable between 1997 and 2006, shrimp farms increase only by about 1300 ha, and the other LULC class decreases by about 1300 ha (Fig. 5.15). The period between 2006 and 2014 mirrors that of the prior period between 1997 and 2006. Mangrove forest cover remains virtually unchanged, whereas shrimp farms increase by a little over 1700 ha and the other LULC class decreases by about 1400 ha. Shrimp farms now occupy far more estuary space than mangrove forests and all other LULC combined.

The 20 years from 1997 to 2007 are a period of rapid and substantial LULC change within the estuaries of El Oro Province. In 1977, shrimp farms occupied only 5% of the study area, whereas mangrove forests occupied 47% and the other LULC class 48% (Fig. 5.16). By 1997, shrimp farms had become the largest LULC in the estuary overtaking both mangrove forest and the other LULC class and occupying more estuarine space than every other LULC class combined. At 51% coverage in 1997, shrimp farms are the dominant landcover in the estuary (Fig. 5.16). From 1997 to 2014, the LULC within the estuaries of El Oro province remains stable, but now in a shrimp farm dominated landscape. Shrimp farms continue to consume more of the study area each decade, but the increases are at a much-reduced rate than the pre-1997 period. Mangrove forests remain stable post-1997 after substantial losses before 1997 (Fig. 5.16). The other LULC class continues to shed area post-1997 but at a much-reduced rate than before 1997. By 2014, it only occupies 19% of the study area whereas, in 1977 it occupied just under half of the study area (Fig. 5.16). The Estuaries of El Oro Province are now better described as shrimp farm environments when basing the description on the long-term dominant LULC.

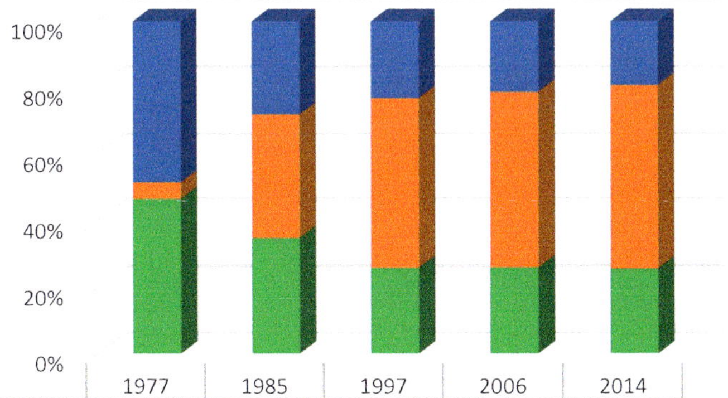

	1977	1985	1997	2006	2014
Other LULC	48%	28%	23%	21%	19%
Shrimp Farms	5%	37%	51%	53%	55%
Mangrove Forest	47%	35%	26%	26%	26%

Fig. 5.16 The estuaries of El Oro province LULC change over time

An additional survey for 2011 is provided for the Estuaries of El Oro Province in addition to the five presented in this analysis, and it can be found in the accompanying Dataverse.

5.2 Landuse and Landcover Displacement Type from the Arrival of Shrimp Aquaculture to 2014 at the Estuarine Level

Across the eight estuaries analyzed, all of them besides Cayapas-Mataje experienced substantial LULC changes during the analysis period. Again, aside from Cayapas-Mataje, all the estuaries' LULC transitioned from mangrove forest dominated environments to shrimp farm dominated environments during the analysis period. This LULC transition occurred during different periods, at different rates, and to various degrees across the eight study sites. These transitions are examined from north to south starting with Cayapas-Mataje Estuary that spans the northern Ecuadorian border into Colombia and ending with the estuaries of El Oro Province that cross the southern border into Peru.

5.2.1 *Cayapas-Mataje Estuary*

Shrimp Farms arrived much later in Cayapas-Mataje Estuary than in the other estuaries to the south, and this is the reason for the initial survey being conducted in 1986. Indeed, many shrimp farms that appeared in the estuary were long gone by the time of the last survey and had reverted to mangrove forest or other LULC. Cayapas-Maatje's LULC remained essentially static between 1986 and 2014 (Fig. 5.17) with no change over almost all of the estuary (Table 5.1). Within REMACAN, only isolated shrimp farms, usually without neighbors, are present and consume little estuarine space when compared to all other estuaries in this study. Indeed, as of 2014, it appears that all the shrimp ponds in REMACAN are abandoned.

In the southwest corner of the study area, outside of REMACAN, a large tract of mangrove forest was lost to shrimp farms and the other LULC class (Fig. 5.17). Some of the mangrove forest gain (Fig. 5.17) seen alongside the interior border of the estuary could be a result of instrument resolution improvements, the accidental inclusion of terrestrial-mangrove mixed forests, or the expansion of mangrove forest possibly related to sea-level rise. More mangrove forest was lost to the other LULC class than to shrimp farms (Table 5.1), but almost all the few small shrimp farms that did arrive are located in areas of former mangrove forest (Table 5.1).

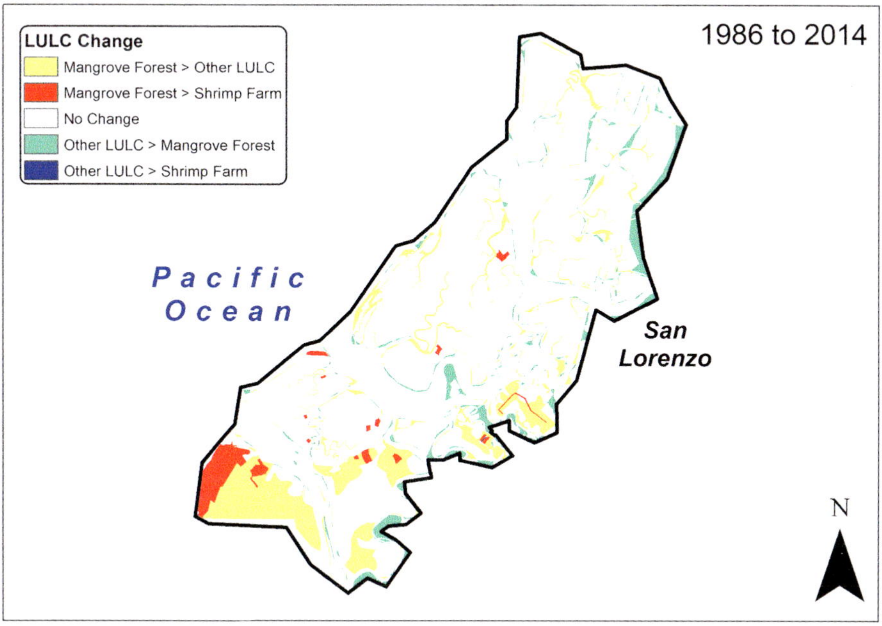

Fig. 5.17 LULC change in Cayapas-Mataje estuary

Table 5.1 LULC change in Cayapas-Mataje Estuary

Output	LULC change	Hectares
0	No change	40,804
-2	Mangrove forest > Shrimp farm	1101
−4	Shrimp farm > Other LULC	Not possible
6	Other LULC > Mangrove forest	2484
−6	Mangrove forest > Other LULC	6234
2	Shrimp farm > Mangrove forest	Not possible
4	Other LULC > Shrimp farm	2

5.2.2 *Muisne Estuary*

In Muisne Estuary, the majority LULC transition from the 1970 survey to the 2014 survey was the change from mangrove forests to shrimp farms (Fig. 5.18). Substantial portions of the inner east side of the estuary were converted from mangrove forest to shrimp farms. Of the 2947 ha of mangrove forest present in Muisne Estuary in the initial survey, 1220 ha are shrimp farms in the final survey (Table 5.2). That is, 41% of the mangrove forest in Muisne Estuary was converted to shrimp farms from 1970 to 2014. In Muisne Estuary, former mangrove forests are the preferred environment for shrimp farms (Fig. 5.18). Areas of mangrove forest were lost to the other LULC class as well.

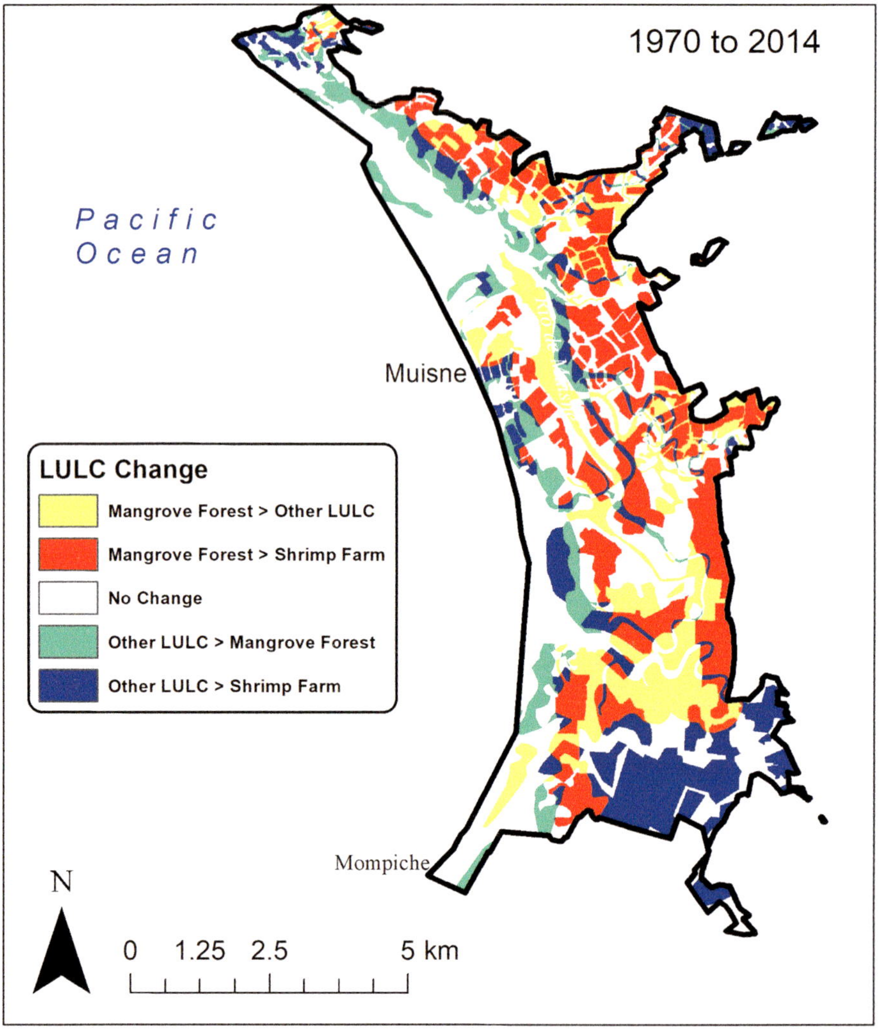

Fig. 5.18 LULC change in Muisne estuary

Table 5.2 LULC change in Muisne estuary

Output	LULC change	Hectares
0	No change	2252
−2	Mangrove forest > Shrimp farm	1220
−4	Shrimp farm > Other LULC	Not possible
6	Other LULC > Mangrove forest	491
−6	Mangrove forest > Other LULC	859
2	Shrimp farm > Mangrove forest	Not possible
4	Other LULC > Shrimp farm	826

Another significant transition within Muisne Estuary is the conversion of the other LULC class to shrimp farms (Table 5.2). Between 1970 and 2014, 826 ha of the other LULC class of mostly surface water and some limited terrestrial environment has become shrimp farms. Not only did mangrove forest convert directly to shrimp farms in Muisne Estuary, but so did substantial amounts of the aquatic environment. In contrast to Cayapas-Mataje Estuary, more of the entire Muisne Estuary's LULC was converted to shrimp farms than remained static over the analysis period (Table 5.2).

5.2.3 *Cojimies Estuary*

In Cojimies Estuary, the major LULC transition from the first survey to the final survey was the LULC change from mangrove forests to shrimp farms (Fig. 5.19). Substantial portions of the inner southeastern and southwestern sections of the estuary converted from mangrove forest to shrimp farm between 1979 and 2014. Of the 13,009 ha of mangrove forest present in Cojimies Estuary in the initial survey (Fig. 5.5), 8364 ha are shrimp farms in the final survey (Table 5.3). That is, 64% of the mangrove forest in Cojimies Estuary was converted to shrimp farms from 1970 to 2014. In Cojimies Estuary, former mangrove forests are the preferred environment for shrimp farms (Fig. 5.19). In Cojimies Estuary, substantial mangrove forest was lost to the other LULC class as well.

The next important transition within Cojimies Estuary is that of Output 4 in Table 5.3. Not only were mangrove forests converted to shrimp farms at a high rate, but the other LULC class was also converted to shrimp farms. Between 1970 and 2014, 3993 ha of the other LULC class became shrimp farms; this other LULC class is mostly surface water and some limited terrestrial environment. Not only did mangrove forest convert directly to shrimp farms in Cojimies Estuary, so did substantial amounts of the aquatic environment. In Cojimies Estuary more of the initial survey LULC was converted to shrimp farms than remained static over the analysis period (Table 5.3).

5.2.4 *Chone Estuary*

In Chone Estuary, the majority LULC transition from the initial survey to the final survey was from mangrove forests to shrimp farming (Fig. 5.20). Substantial portions of the northern and the inner estuary have converted from mangrove forest to shrimp farms between 1968 and 2014. Of the 4207 ha of mangrove forest present in Chone Estuary in the initial survey (Fig. 5.7), 2590 ha are shrimp farms in the final survey (Table 5.4). That is, 62% of the mangrove forest in Chone Estuary was converted to shrimp farming from 1968 to 2014. In Chone Estuary, former mangrove forests are the preferred environment for shrimp farms.

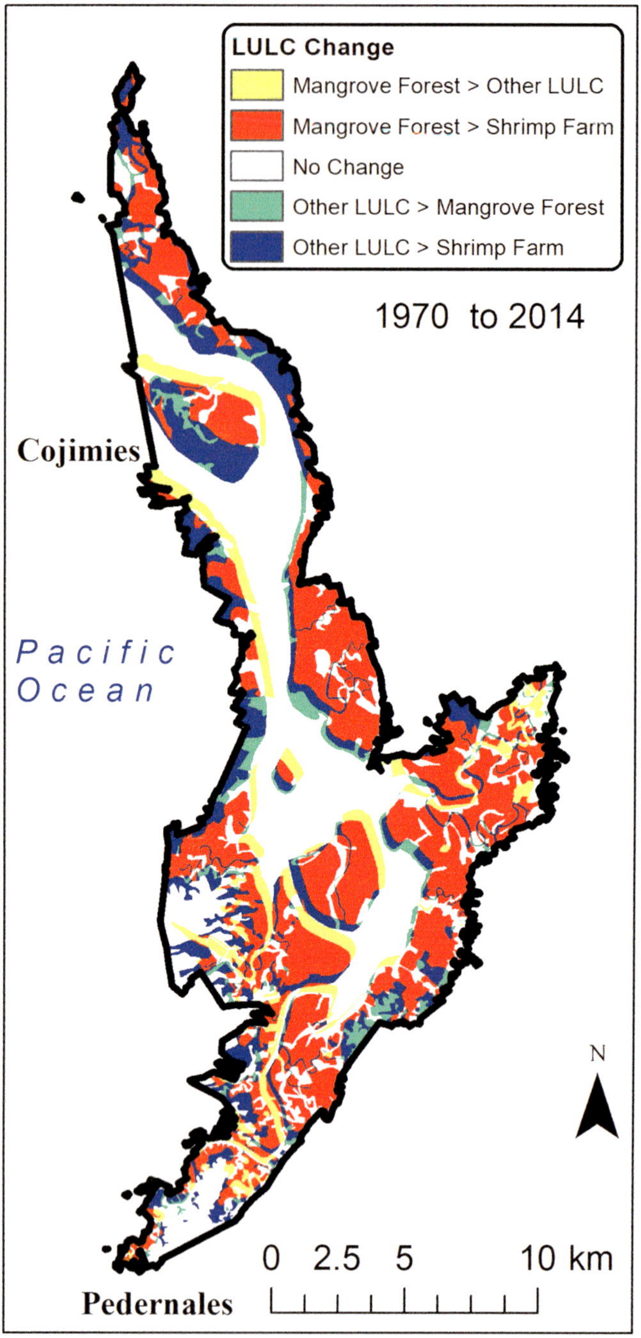

Fig. 5.19 LULC change in Cojimies estuary

Table 5.3 LULC change in Cojimies estuary

Output	LULC change	Hectares
0	No change	9573
−2	Mangrove forest > Shrimp farm	8364
−4	Shrimp farm > Other LULC	Not possible
6	Other LULC > Mangrove forest	1374
−6	Mangrove forest > Other LULC	2088
2	Shrimp farm > Mangrove forest	Not possible
4	Other LULC > shrimp farm	3993

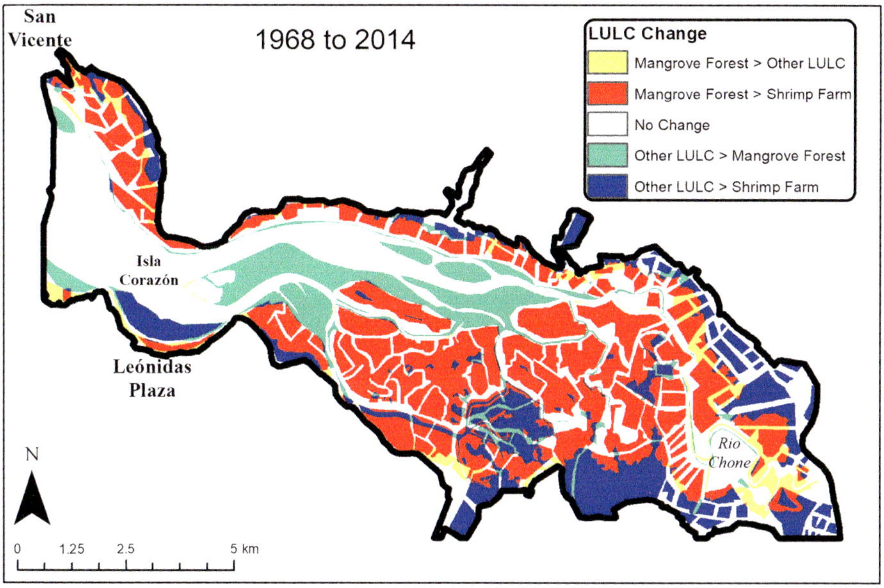

Fig. 5.20 LULC change in Cojimies estuary

The second important transition to note is that of Output 4 in Table 5.4. Not only were mangrove forests converted to shrimp farms at a high rate, but so was the other LULC class. Between 1968 and 2014, 1269 ha of the other LULC class became shrimp farms. Like Muisne Estuary, this is mostly surface water and some limited terrestrial environment. Thus, not only did mangrove forest convert directly so shrimp farms in Chone Estuary as did substantial amounts of the aquatic environment. The final transition of note in Fig. 5.20 is the other LULC conversion to mangrove forest (Table 5.4). Almost all the new mangrove forest that appeared in the post−2000 analysis period (Fig. 5.7) is in areas that never traditionally had mangrove forest but were instead part of the aquatic environment (Fig. 5.20). In Chone Estuary, more of the estuary converted to shrimp farms than remained static (Fig. 5.20).

Table 5.4 LULC change in Chone estuary

Output	LULC change	Hectares
0	No change	3470
−2	Mangrove forest > Shrimp farm	2590
−4	Shrimp farm > Other LULC	Not possible
6	Other LULC > Mangrove forest	937
−6	Mangrove forest > Other LULC	389
2	Shrimp farm > Mangrove forest	Not possible
4	Other LULC > Shrimp farm	1269

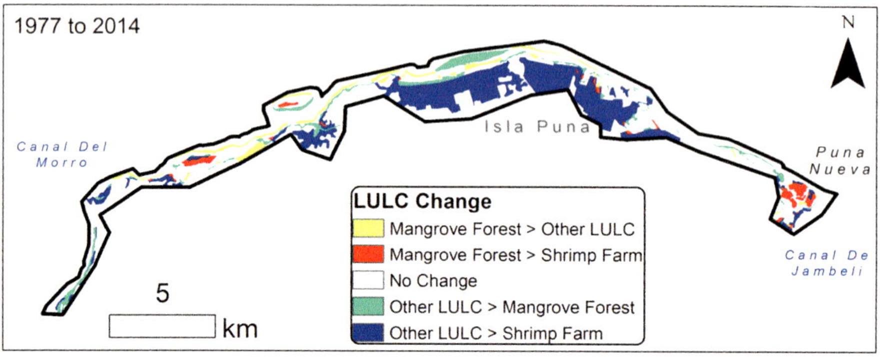

Fig. 5.21 LULC change in Isla Puná North

5.2.5 *Isla Puná North*

The LULC transition on Isla Puná North does not follow the same pattern as the estuaries of Chone, Cojimies, or Muisne further north. On Isla Puná North the only substantial LULC change is from the other LULC class to shrimp farms (Fig. 5.21). Between 1977 and 2014 the other LULC class, consisting mostly of water, salt-pans, and terrestrial land in this study area, had 1830 ha (Table 5.5) of their area converted to shrimp farms. This equates to 32% of the initial survey's other LULC class converting to shrimp farms (Fig. 5.9). Mangrove forest did not convert to shrimp farms at any meaningful level on Isla Puná North; indeed more mangrove forest was lost to other LULC than to shrimp farming. Isla Puná North mangrove forests were not the preferred environment of shrimp farms once they arrived in the area (Table 5.5).

Table 5.5 LULC change on Isla Puná North

Output	LULC change	Hectares
0	No change	4343
−2	Mangrove forest > Shrimp farm	184
−4	Shrimp farm > Other LULC	Not possible
6	Other LULC > Mangrove forest	441
−6	Mangrove forest > Other LULC	353
2	Shrimp farm > Mangrove forest	Not possible
4	Other LULC > Shrimp farm	1830

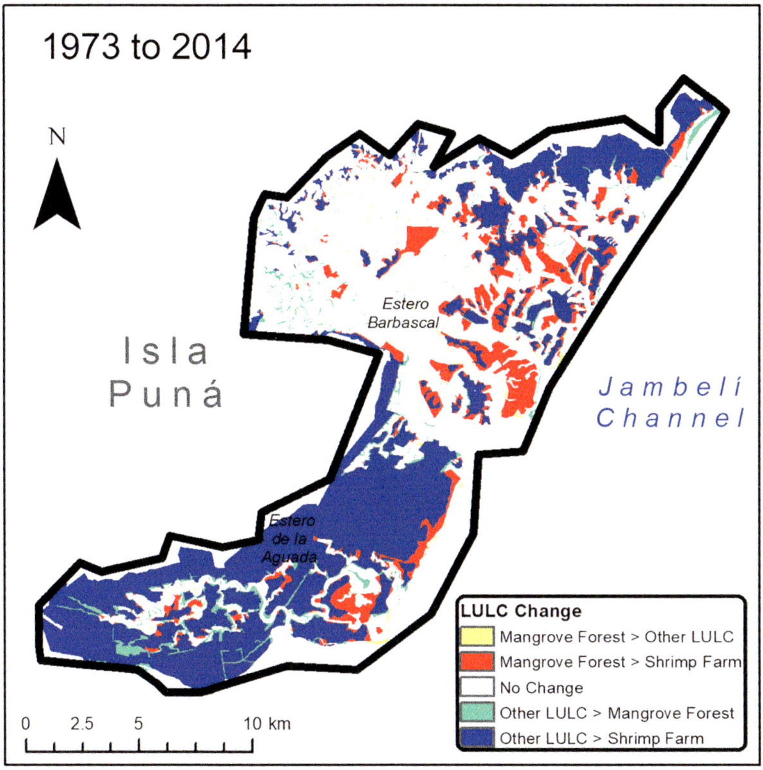

Fig. 5.22 LULC change on Isla Puná

Table 5.6 LULC change on Isla Puná

Output	LULC change	Hectares
0	No change	16,436
−2	Mangrove forest > Shrimp farm	2676
−4	Shrimp farm > Other LULC	Not possible
6	Other LULC > Mangrove Forest	1243
−6	Mangrove forest > Other LULC	361
2	Shrimp farm > Mangrove forest	Not possible
4	Other LULC > Shrimp farm	10,525

5.2.6 *Isla Puná*

The LULC transition on Isla Puná does not follow the same pattern as the estuaries of Chone, Cojimies, or Muisne further north but does follow the LULC change pattern of neighboring Isla Puná North. On Isla Puná, the largest LULC change is from the other LULC class to shrimp farms (Fig. 5.22). Between 1977 and 2014, the other LULC class, consisting mostly of water, salt-pans, and terrestrial land in this study area, had 10,525 ha (Table 5.6) of its initial area converted to shrimp farms. Table 5.6 reports that just under 50% of the initial survey's other LULC class converted to shrimp farms (Fig. 5.11). Unlike Isla Puná North, Isla Puná also has noteworthy mangrove forest conversion to shrimp ponds with 2676 ha of the initial survey's 9947 ha of mangrove forests converted to shrimp ponds by 2014 (Table 5.6). In Isla Puná mangrove forests were not the preferred environment of shrimp farms once they arrived in the area, the other LULC class was preferred instead (Table 5.6).

5.2.7 *Guayas Estuary*

Determining the LULC transitions in Guayas Estuary is difficult as over 30,000 ha of shrimp farms existed in the estuary before the initial Landsat survey. For this reason, the delineation of comprehensive LULC for Guayas estuary in the 1970s remains a research goal. Fig. 5.23 depicts the LULC transition, but the estuary had already entered somewhat of a stable LULC domain by this period. Additionally, the large size of the estuary makes it difficult to see the more micro LULC transitions. Despite these caveats, between 1985 and 2014 large portions of the eastern estuary converted from mangrove forest to shrimp farms (Fig. 5.23). The amount of mangrove forest lost directly to shrimp farms totals 21,565 ha and is the largest single LULC transition in the estuary (Table 5.7).

On the western side of the estuary, where shrimp farms were already dominant, the primary LULC conversion is from the other LULC class to shrimp farms (Fig. 5.23). Over the entirety of Guayas Estuary, the other-LULC to shrimp farms is the second largest transition after 1985 (Table 5.7). One benefit of having over 30,000 ha of shrimp farms present in the earliest survey is the conversion from

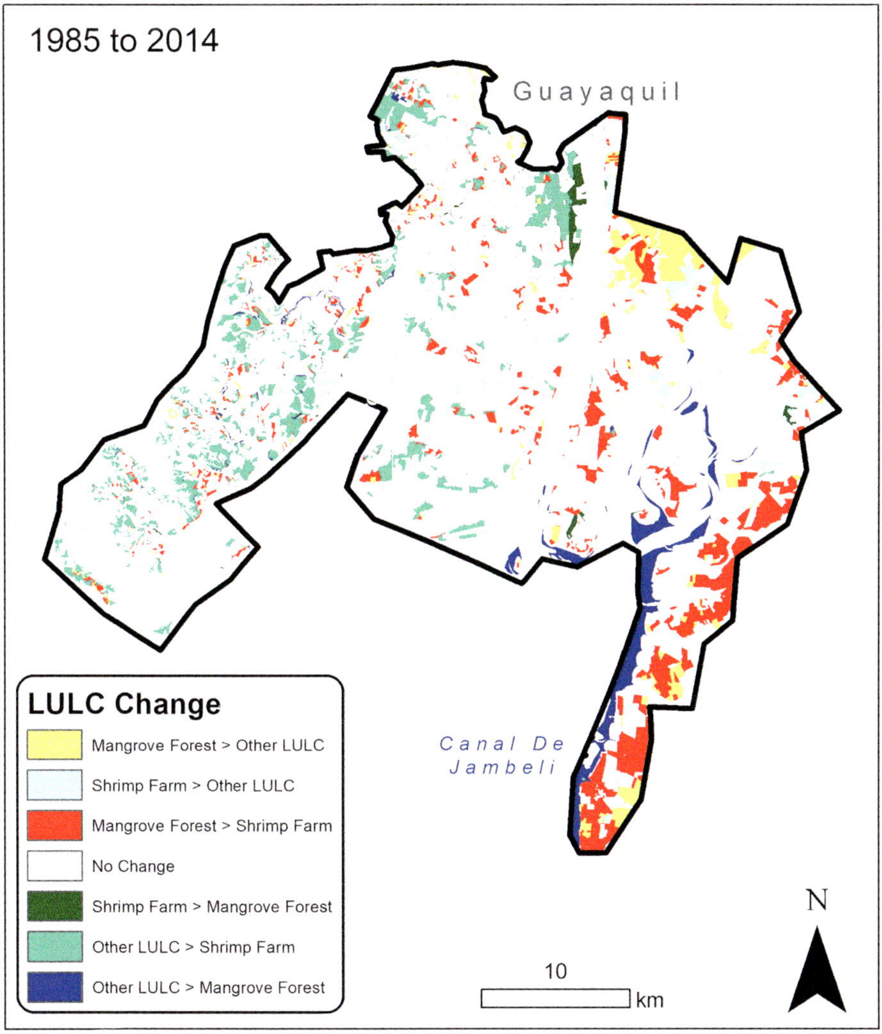

Fig. 5.23 LULC change in Guayas estuary

Table 5.7 LULC change in Guayas estuary

Output	LULC change	Hectares
0	No change	227,578
−2	Mangrove forest > Shrimp farm	21,565
−4	Shrimp farm > Other LULC	3541
6	Other LULC > Mangrove forest	8355
−6	Mangrove forest > Other LULC	10,133
2	Shrimp farm > Mangrove forest	1409
4	Other LULC > Shrimp farm	17,393

shrimp farm LULC to another LULC can be monitored. Of the land that was under shrimp farming in 1985, only a nominal amount has converted to mangrove forest or the other LULC class (Table 5.7). One final item to note is that the new mangrove forest that has appeared is primarily in areas of the other LULC class and not in the shrimp farm LULC. That is, new mangrove forests primarily appear in areas that traditionally lacked mangrove forest.

5.2.8 *The Estuaries of El Oro Province*

Within the estuaries of El Oro Province, shrimp farms have directly replaced mangrove forest cover across much of the estuary (Fig. 5.24) with 16,933 ha displacing mangrove forest between 1977 and 2014 (Table 5.8). That is, 58% of all the mangrove forests detected in the original survey within the estuaries of El Oro province are now shrimp farms. It is not only mangrove forest that was directly displaced by shrimp farms in the Estuaries of El Oro Province but also the other LULC class. Indeed, the other LULC class appeared to be the preferred environment to locate shrimp farms in the estuaries of El Oro province, as other LULC to shrimp farm is the single largest LULC transition in the study area closely followed by mangrove forests to shrimp farms (Table 5.8). The combined total of mangrove forest to shrimp farm conversion in the estuary is 35,439 ha, and this is greater than 50% more than the amount of LULC that remained static in the estuary. That is, far more of the estuary converted to estuary than did not change LULC.

5.3 Landuse and Landcover Change from the Arrival of Shrimp Aquaculture to 2014 at the Country Level

Across all the study sites in coastal Ecuador, 232,857 ha of mangrove forests were present in the initial survey, meaning mangrove forests constituted 48% of all estuarine space (Fig. 5.25). The other LULC class, consisting mostly of water covered, 220,743 ha or 45% of all estuarine space at the first survey. (Figure 5.25) Finally, 7% of estuarine space or 34,252 ha was made up of shrimp farms, existing only in the estuaries of El Oro Province and Guayas Estuary (Fig. 5.25) where pre-shrimp farming LULC could not be obtained. The total study area size is 487,853 ha.

Across all the study sites in coastal Ecuador, 176,937 ha of mangrove forest remained in the final survey of 2014. The mangrove forests constituted 36% of all estuarine space in the final survey (Fig. 5.26). The other LULC class, consisting mostly of water, covered 173,254 ha or 36% of all estuarine space in the 2014 survey (Fig. 5.26). Finally, 28% of estuarine space or 137,662 ha was made up of shrimp farms (Fig. 5.26). It is worth noting that in all estuaries aside from Cayapas-Mataje, Isla Puná North, and Guayas that shrimp farms are the dominant LULC. It

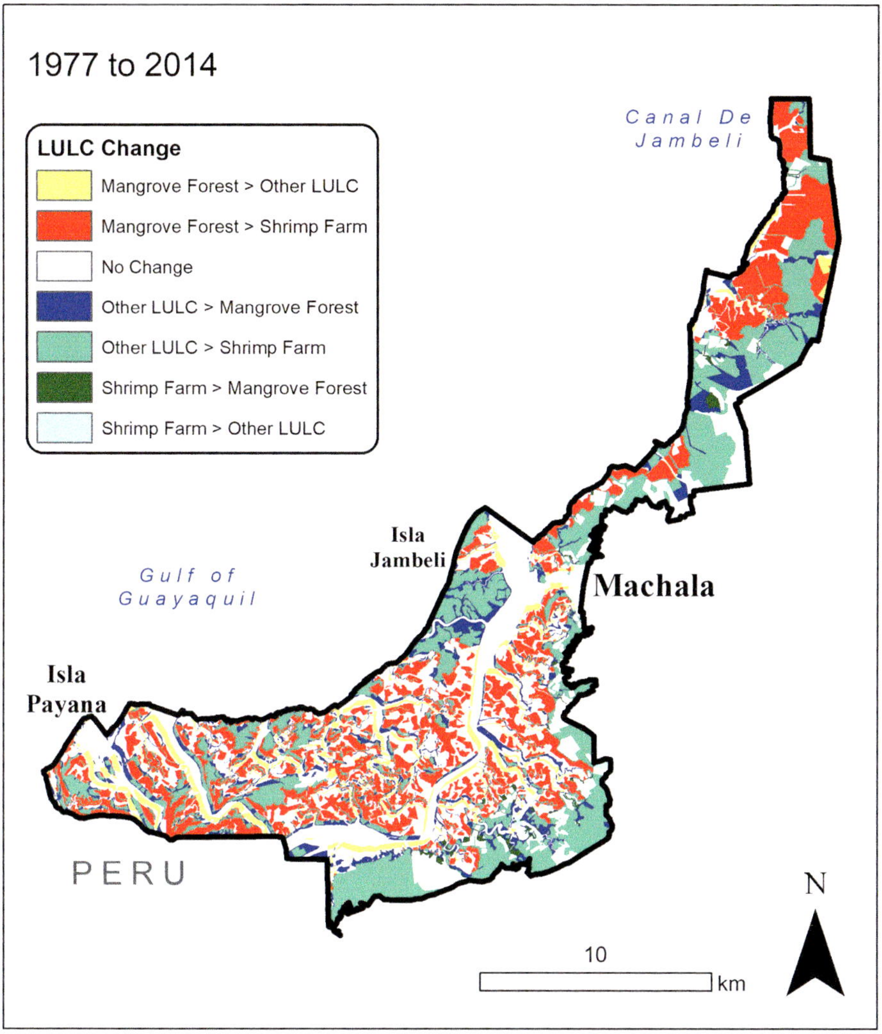

Fig. 5.24 LULC change among the estuaries of El Oro province

Table 5.8 LULC change among the estuaries of El Oro province

Output	LULC change	Hectares
0	No change	22,357
−2	Mangrove forest > Shrimp farm	16,933
−4	Shrimp farm > Other LULC	187
6	Other LULC > Mangrove forest	6325
−6	Mangrove forest > Other LULC	4366
2	Shrimp farm > Mangrove forest	450
4	Other LULC > Shrimp farm	18,506

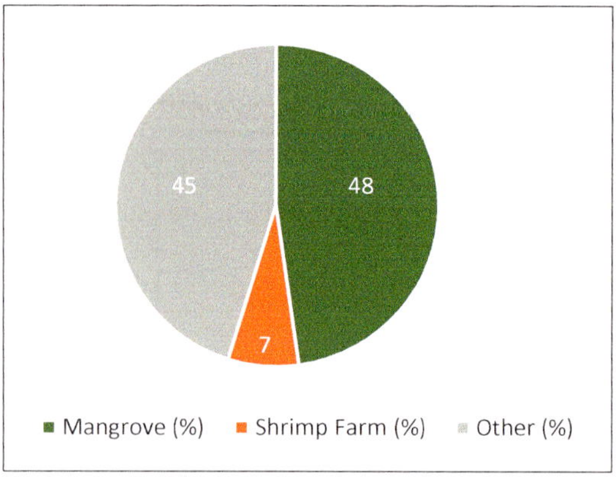

Fig. 5.25 Initial survey LULC across all estuaries

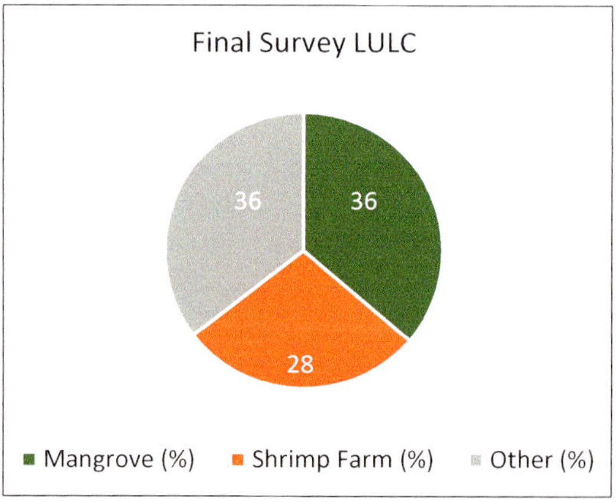

Fig. 5.26 Final 2014 survey LULC across all estuaries

is also worth noting that when presented without the REMACAN site of Cayapas-Mataje, the final survey percentages change to 34% mangrove forest, 35% other LULC class, and 31% for shrimp farms, resulting in an almost equal split across the three LULC classes. That is, excluding the REMACAN site of Cayapas-Mataje, the LULC change would have been more substantial. Excluding both Cayapas-Mataje Estuary and Guayas Estuary, the shrimp farm percentage jumps up to 49% of all estuarine space in the final 2014 survey.

Three distinct periods of mangrove forest loss are identified in these data. First is an early period of slight mangrove forest loss when shrimp farms initially arrived in the estuary. The second identified period is of rapid and substantial mangrove forest loss in the decade following the first arrival of shrimp farms. The final period is defined by relative stability or even a slight recovery of mangrove forest, albeit from a much-diminished level due to earlier losses. The three periods roughly correspond to the 1970s and early 1980s, the mid-1980s and 1990s, and post-2000 respectively, although this is somewhat simplified and does not account for the late arrival in the 1980s of shrimp aquaculture in Cayapas-Mataje. By 2014, 176,937 ha of the initial survey level of mangrove forest remained across all of Ecuador, but this had dropped to as low as 170,639 ha around the year 2000. Hence the final survey represents somewhat of a recovery from the point of the lowest mangrove forest area at the turn of the century.

The two exceptions to the Ecuadorian-level mangrove forest loss trends are Cayapas-Mataje and much of Isla Puná. The Isla Puná exception is that mangrove forest was not substantially lost, although the other LULC class did lose substantial area to shrimp farms. The Isla Puná exception likely relates to either the remote island location of the mangrove forests and difficulty of clearing them via mechanized means or a period of clearing before the shrimp farms arrived that is captured in the initial imagery. That is, the imagery captures the cleared areas of mangrove forest just before the first aquaculture ponds arrived, and hence it appears that a loss of the other LULC class occurred when the same area was likely mangrove forest a few months or years earlier. Unfortunately, this proposed earlier LULC transition cannot be confirmed and is likely lost to history.

During the initial surveys, 34,252 ha of shrimp farms were in the combined study areas. By the final surveys in 2014, shrimp farms consumed 173,254 ha of all of Ecuador's estuarine space. This expansion of shrimp farms from what would have been close to 0 ha in 1970 to 173,254 ha in 2014 is the single greatest LULC change along coastal Ecuador during the study period.

Within the estuaries of Cojimies, Chone, and El Oro Province the combined LULC transition to shrimp farms is greater than the amount of those study areas that remained static, and this threshold was almost surpassed in Muisne Estuary. That is, more estuarine area was converted to shrimp farms than that did not convert or had a different LULC conversion. This remarkable statistic demonstrates the rapid and substantial LULC change that occurred across most of Ecuador after the arrival of shrimp aquaculture. Where mangrove forests once dominated, only some limited mangrove forests now remain in shrimp-farmed estuaries.

5.4 Landuse and Landcover Displacement Type from the Arrival of Shrimp Aquaculture to 2014 at the Country Level

Shrimp farms did not only expand in extent from their first arrival in the estuaries of Ecuador but they also directly displaced mangrove forest cover at substantial levels across most estuaries. The preferred LULC class for shrimp farm locations was in mangrove forests in all estuaries aside from Isla Puná, Cayapas-Mataje, and El Oro Province. Within the estuaries of El Oro Province this criterion would likely have been met if early shrimp farms' initial LULC could be determined, and on Isla Puná the criterion is possibly met but is not captured by the data (Table 5.9). It is only on Isla Puná, across both estuaries, that minor amounts of mangrove forest were converted to shrimp farms and the other LULC class was preferred for shrimp farm

Table 5.9 Preferred initial LULC for shrimp farms

Study area	Preferred initial LULC	Percentage displaced by shrimp farms
(i) Cayapas-Mataje Estuary	Mangrove	100
	Non-mangrove	0
(ii) Muisne Estuary	Mangrove	60
	Non-mangrove	40
(iii) Cojimies Estuary	Mangrove	77
	Non-mangrove	23
(iv) Chone Estuary	Mangrove	67
	Non-mangrove	33
(vi) Isla Puná North	Mangrove	9
	Non-mangrove	91
(vii) Isla Puná	Mangrove	20
	Non-mangrove	80
(vii) Guayas Estuary	Mangrove	55
	Non-mangrove	45
(viii) Estuaries of El Oro	Mangrove	48
	Non-mangrove	52

expansion. Across the northern estuaries, aside from the long-term protected Cayapas-Mataje Estuary, the amount of mangrove forest directly displaced by shrimp farms is substantial.

In Guayas Estuary, Chone Estuary, Muisne Estuary, Cojimies Estuary, and Cayapas-Mataje Estuary, the preferred location for shrimp farms is mangrove forest by a substantial margin. In El Oro, the other LULC is the preferred habitat of shrimp farms but by a slight amount. On Isla Puná the other LULC locations are preferred. It should be noted that Isla Puná is a small area compared to other estuaries and that Cayapas-Mataje only ever had a handful of shrimp farms. The data provided here for Ecuador indicate that, with the odd exception, the most favored location for shrimp farms within Ecuadorian estuaries was former mangrove forests, even when other suitable locations were available.

5.5 Patterns of Shrimp Farm Expansion and Mangrove Forest Loss at the Sub-Estuarine Level: An Issue of Property Rights Enforcement

The spatiotemporal pattern of shrimp farm expansion is best revealed in Chone Estuary with semi-decadal surveys. A distinct pattern of shrimp farm expansion is clear. During the initial period of shrimp farm arrival in the estuary around 1970, the ponds displaced mangrove forests in the estuary, and the individual ponds were larger in surface area when compared to shrimp ponds that arrived later. The author verified that these early farms were and remain extensive in nature and require little or no feed or other inputs for the shrimp to grow into mature market-ready shrimp. After the initial arrival of extensive shrimp farms in Chone Estuary that likely occupy the prime locations within the estuary, the remaining farms that arrive are all semi-intensive operations. These semi-intensive growout ponds constructed in Chone Estuary in the 1980s and 1990s are smaller in size, far more numerous, and generally occupy both former mangrove forest or estuary waters along the terrestrial edge of the estuary. Particularly along the northern boundary of the estuary, the shrimp growout ponds appear in linear strips along the estuary edge initially, then over the next 20 years they extend out farther into the estuary.

During public participation geographic information systems (PPGIS) workshops hosted in Chone Estuary in 2008, members of the local fishing syndicate were asked, "Why did you sell your land to the shrimp farm companies?" The responses almost universally revealed that they were tricked into selling their land and never anticipated what would follow. The fishers explained that they sold a small plot of land, typically tens of hectares on the estuarine edge, to shrimp farm developers at a fair price. Developers would develop shrimp farms on these parcels of land and extend them out into the adjacent estuary, which the developers did not own. Then over subsequent years, new shrimp ponds would be constructed ever further out into the estuary but also parallel to the shoreline. This expansion of shrimp ponds over time would limit the adjacent properties' access to the estuary and mangrove forests. The traditional fishing communities still reside on these adjacent properties today.

Figure 5.27 is a diagram representing how selling a single shoreline property on the northern shoreline of Chone Estuary resulted in all adjacent properties losing access to both the mangrove forest and the wider estuary. In this example, property 3 is sold to the shrimp farm company or individual. The y-axis across all panels is time, from approximately 1980 in the top panel to approximately 2000 in the bottom panel. The timestep is 5 years. The LULC is depicted across all panels' x-axes and y-axes. This diagrammatic representation of the spatiotemporal dimension of the LULC data for Chone Estuary (Fig. 5.27) supports the land grab oral history as explained by long-term residents of the estuary in the PPGIS session. That is, the remote sensing data supports the testimony of the traditional fishers who attended the PPGIS session and claimed that shrimp farms expanded from limited terrestrial purchases.

In Panel 1 (Fig. 5.27), the traditional fishing community owns all land along the coastline and has full access to the estuary and mangrove forest. By Panel 2, the shrimp farms are built but primarily located on the purchased land and just offshore into the communal mangrove forest and estuary. By Panel 3, the shrimp farms have expanded out not only into the communal mangrove forest and into the wider estuary but have also expanded parallel to the shoreline and begun to limit other properties' access to the mangrove forest and estuarine resources. In the final Panel 4, the shrimp farm expansion is complete, and the pre-aquaculture fishing community has lost access to all the mangrove forest and the wider estuary as shrimp farm expansion has continued both out into the estuary and parallel to the shoreline. Fig. 5.27

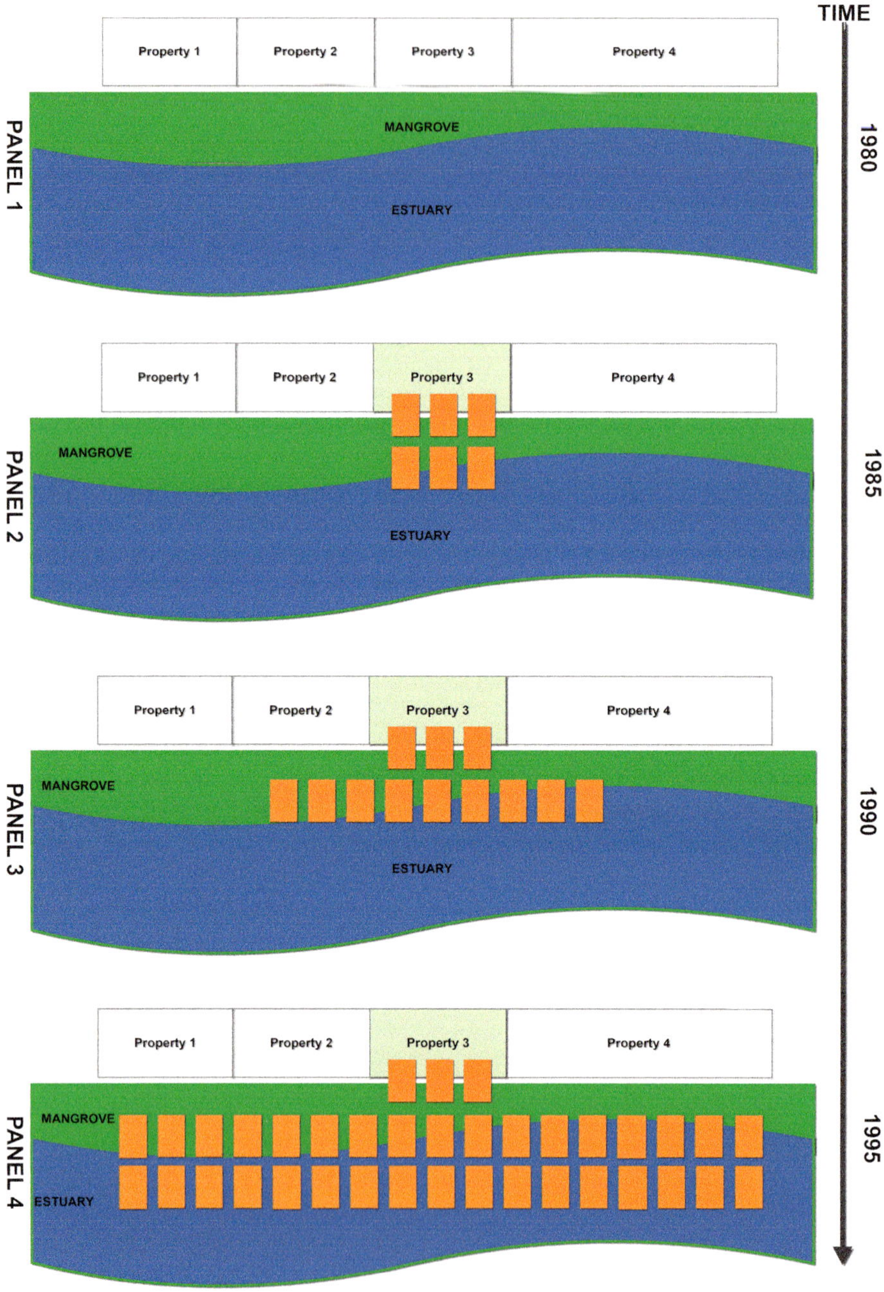

Fig. 5.27 Shrimp farm expansion and property rights enforcement

demonstrates how shrimp farm expansion could result in the shrimp farm owner having many hundreds of hectares of ponds while only purchasing a few hectares of terrestrial land from the original fishing community. In addition, Fig. 5.27 demonstrates how shrimp farm expansion over time limited traditional fishing communities' access to both the mangrove forest and estuarine resources.

5.6 Results Summary

Mangrove forest loss occurred across all estuaries in Ecuador from the arrival of shrimp aquaculture to 2014. Mangrove forest loss was substantial across all estuaries aside from Isla Puná, which is the smallest estuary and has an almost insignificant percentage of the entire country's mangrove forest. In the estuaries of Muisne, Cojimies, Chone, and El Oro, mangrove forest losses are substantial, and some of the northern estuaries now only have a small fraction of their initial mangrove forest remaining. Much of the mangrove forest loss occurred within a decade of the first arrival of aquaculture in an estuary, and mangrove deforestation generally continued but at reduced rates until about 2000. Mangrove forests are now the majority estuarine landcover only in Cayapas-Mataje Estuary, whereas in the initial survey, mangrove forests were the dominant landcover in at least four of the estuaries. Although mangrove forest loss has been substantial since the advent of commercial aquaculture, there has been limited mangrove forest recovery, albeit at a low level, since about 2000. Future surveys will be required to determine if this trend of mangrove forest recovery is a long-term pattern or merely a short-term phenomenon.

Substantial shrimp farm expansion has occurred across all estuaries aside from Cayapas-Mataje Estuary in which shrimp farms only occupy 2% of the estuarine space. In four of the eight estuaries, shrimp farms are now the dominant LULC in the estuary with shrimp farms occupying more space than either mangrove forest or the other LULC class that includes surface water. Although the shrimp farm expansion occurred in all estuaries, it occurred during different periods and to different extents in each estuary. Once the initial shrimp ponds arrived in the estuary, the estuary appeared to rapidly transition from mangrove forest or other LULC dominated to shrimp farm dominated. As of 2014, shrimp farms are still being constructed in Ecuador's estuarine environments, although unlike the earlier shrimp farms, the new farms appear to be constructed on the estuary edge as opposed to their earlier more central estuarine aquatic location.

There is no evidence of meaningful declines in shrimp farm extent in any of Ecuador's estuaries during any period included in this analysis. That is, shrimp farms, even when abandoned, do not appear to have reverted to their initial LULC in any estuary. As shrimp farms arrived in the estuary, the preferred location was in the mangrove forests. In five of the eight estuaries, shrimp farms replaced mangrove forest habitat; in one estuary the preferred location was about equal between mangrove forest and the other LULC class. Not only were mangrove forests were replaced, but the other LULC class was also replaced by shrimp farms once they arrived. Indeed, on Isla Puná the shrimp farm expansion appears to favor the other LULC class. This study does not attempt to quantify the level of shrimp farms outside of Ecuador's major estuaries and hence cannot be used to estimate the number of shrimp ponds in the entire country. However, the results indicate that the introduction of shrimp farming in Ecuador had a major impact on LULC in all of Ecuador's estuaries and primary mangrove habitat.

Data Availability

All data is stored in a Harvard Dataverse at https://dataverse.harvard.edu/dataverse/MFACRL

Cayapas-Mataje Estuary final classified LULC analysis data for each year and supplemental years as noted (all of Fig. 5.1) is available at https://doi.org/10.7910/DVN/7DB68C

Muisne Estuary final classified LULC analysis data for each year (all of Fig. 5.3) is available at https://doi.org/10.7910/DVN/7DB68C

Cojimies Estuary final classified LULC analysis data for each year (all of Fig. 5.5) is available at https://doi.org/10.7910/DVN/7DB68C

Chone Estuary final classified LULC analysis data for each year (all of Fig. 5.7) is available at https://doi.org/10.7910/DVN/7DB68C

Isla Puná North Estuary classified LULC analysis data for each year (all of Fig. 5.11) is available at https://doi.org/10.7910/DVN/7DB68C

Isla Puná Estuary classified LULC analysis data for each year (all of Fig. 5.13) are available at https://doi.org/10.7910/DVN/7DB68C

Guayas Estuary classified LULC analysis data for each year (all of Fig. 5.15) are available at https://doi.org/10.7910/DVN/7DB68C

El Oro Estuary classified LULC analysis data for each year (all of Fig. 5.17) is available at https://doi.org/10.7910/DVN/7DB68C

Cayapas-Mataje Estuary final classified LULC-change analysis data for each year and supplemental years as noted (all of Fig. 5.17) is available at https://doi.org/10.7910/DVN/482JFM

Muisne Estuary final classified LULC-change analysis data for each year (all of Fig. 5.18) is available at https://doi.org/10.7910/DVN/482JFM

Cojimies Estuary final classified LULC-change analysis data for each year (all of Fig. 5.19) are available at https://doi.org/10.7910/DVN/482JFM

Chone Estuary final classified LULC-change analysis data for each year (all of Fig. 5.21) is available at https://doi.org/10.7910/DVN/482JFM

Isla Puná North Estuary classified LULC-change analysis data for each year (all of Fig. 5.22) is available at https://doi.org/10.7910/DVN/482JFM

Isla Puná Estuary classified LULC-change analysis data for each year (all of Fig. 5.23) is available at https://doi.org/10.7910/DVN/482JFM

Guayas Estuary classified LULC-change analysis data for each year (all of Fig. 5.24) is available at https://doi.org/10.7910/DVN/482JFM

El Oro Estuary classified LULC-change analysis data for each year (all of Fig. 5.17) is available at https://doi.org/10.7910/DVN/482JFM

Initial Survey Master Excel (Fig. 5.25) is available at https://doi.org/10.7910/DVN/IVP3RD

Final Survey Master Excel (Fig. 5.26) is available at https://doi.org/10.7910/DVN/IVP3RD

Chapter 6
Discussion: Local Actions with Global Consequences

Abstract This chapter discusses the results obtained in this research with a focus on the national scale and attempts to fit this analysis into wider research themes based around commodification of land use, livelihood change due to LULC-change, and the human-dimensions of LULC-change. Following the results summary, the chapter discusses a period of recent mangrove forest expansion revealed in the analysis and expands upon the potential reason for this mangrove forest increase and the future monitoring of this increase that will be required to ascertain if it is an actual trend or merely a data anomaly. The chapter then raises the possibility of a potential convergence of numerous factors at differing scales that likely offer the potential to reverse the decades-long clearing of mangrove forests that are revealed in this analysis. Next, the chapter discusses how and why Cayapas-Mataje managed to preserve its mangrove forest in the face of such substantial and rapid declines elsewhere along the coast of Ecuador. Finally, some policy recommendations and forward-looking methodological observations are made to close out the discussion.

Keywords Deforestation · Reforestation · Mangrove loss · International financing

6.1 The Ecuadorian Estuarine Mangrove Loss Story. The Commodification of Ecuador's Estuarine Environments

The results of this long-term remote sensing analysis have demonstrated that Ecuador's estuaries have transitioned from a mangrove forest dominated landscape to a landscape that is now dominated by commercial shrimp aquaculture growout pond. Certain estuaries such as Chone Estuary and Cojimies Estuary have had a near total conversion of their initial mangrove forest to shrimp farms while in other estuaries such as those of Isla Puná and Cayapas-Mataje the conversion from mangrove forest to shrimp farm has still occurred but in a more limited manner. It is

S. E. Hamilton, *Mangroves and Aquaculture*, Coastal Research Library 33,
https://doi.org/10.1007/978-3-030-22240-6_6

likely that the transition from mangrove forest to shrimp farms during the analysis period is even more dramatic than presented in this manuscript. The reason for this potential underestimation of mangrove forest loss is that the initial LULC surveys in Guayas and El Oro provinces combined contained approximately 35,000 ha of shrimp farms for which the pre-aquaculture LULC could not be determined. It can be extrapolated from the other estuaries and the earlier data we have for El Oro that the majority, if not nearly all, of these early shrimp farms likely occupied earlier mangrove forests. Even if only 50% of these shrimp farms resided in former mangrove forest, this would result in the amount of mangrove lost to shrimp farm being equivalent to the total amount of mangrove forests present in the initial surveys for all of Muisne Estuary, Chone Estuary, and Isla Puná North combined. That is, the amount of shrimp farm to mangrove conversion although remarkable as presented is likely even higher than reported.

The results section of this papers reveals that the expansion of commercial shrimp aquaculture in Ecuador's estuaries was both rapid and massive. From what would have been 0 ha, or at least close to 0 ha in 1970 (or from 34,252 ha at the initial survey) shrimp farms expanded to 173,254 ha by 2014. If it were not for Cayapas-Mataje shrimp farms would consume as much estuarine space as mangrove forest in Ecuador. Even including the near pristine Cayapas-Mataje site, shrimp growout ponds still occupy approximately 28% of all estuarine space compared to 36% for mangrove forests, but at the local scale, the conversion is even more remarkable. Shrimp farms now occupy more estuarine space than mangrove forests in the six of the eight estuaries analyzed and consume more estuarine space than either mangrove forest or everything other LULC-class in half of the estuaries. Indeed, only Cayapas-Mataje Estuary and Guayas Estuary have mangrove forest remaining as their dominant LULC.

This analysis reveals that shrimp farms preferred location for construction in Ecuador was mangrove forest. In five of the eight estuaries, shrimp farms preferred to locate in mangrove forest even when other LULC-classes were available, and in one of the others, it was almost even between preferring mangrove forest or all other LULC-classes. Even when shrimp farms preferred the location was in other LULC-classes, shrimp farm still decimated the mangrove forests and did not expand in these other regions as opposed to expanding mangrove forests; they merely expanded in both LULC-types. That is, shrimp farm expansion in other areas of the aquatic environment did not result in much if any mangrove preservation. Indeed, this remote sensing analysis reveals that greater than 90% of all mangrove forest loss across mainland Ecuador was due to the direct displacement of mangrove forest by shrimp farms. Ecuadorian estuaries do not appear to follow the often-repeated global pattern of aquacultures preferred location being outside of mangrove forest. Indeed it has been stated that less than 10% of global shrimp ponds are located in former mangrove forests with a maximum of 20% in some countries (Boyd and Clay 1998) although this statement is challenged (Hamilton 2015). It is evident that in Ecuador over this analysis period that shrimp farms are directly and almost exclusively responsible for all the mangrove forest loss that has occurred.

In many regards, the estuaries of Ecuador are now no different than any other commodity-driven agricultural landscape such as a cornfield in a farm Iowa USA, a rice field in a farm in Yunnan China, or a sugarcane field in a farm in South-Central Brazil. The primary LULC of coastal Ecuador is now dedicated to a single species crop; in this case the crop is Whiteleg Shrimp. Like many other crops, it is harvested a couple of times per year, it requires feed and other inputs to survive and grow, and when harvested almost all the crop is exported to developed markets for consumption. What is remarkable and is revealed in the remote sensing always is the pace and level of the shrimp farm expansion in Ecuador, from being at a nominal level in 1980 shrimp farms expanded into a position of complete landscape dominance by around 2000. Ecuador's estuaries can be stated to have undergone a rapid and near complete commodification over approximately 20-years in most locations.

6.2 The Importance of Long-Term and High-Resolution Mangrove Monitoring

The importance of the temporal resolution of this analysis is paramount. Few if any mangrove loss analyses report data from before the year 2000. The year 2000 was when researchers delivered MFW, MFW is a high-resolution global coverage map of mangrove forests (Giri et al. 2011) applicable in Ecuador and beyond and is used in hundreds of studies to establish mangrove forest cover baselines. If this survey had taken the approach of basing the start date on the ease of data availability made available due to products such as MFW, GFC (Hansen et al. 2013), or CGMFC-21 (Hamilton and Casey 2016) then the results would look much different and tell a wholly opposite story from the narrative presented in this manuscript. Mangroves would appear to be gaining in area, whereas; shrimp farms would be almost static in area or growing moderately. The survey dates chosen by the researcher would purely drive this LULC-pattern as opposed to the longer-term reality on the ground. This manuscript demonstrates the importance of avoiding arbitrary start dates in a LULC-change analysis and using event-driven or other dates that are meaningful on the ground. In this case, the triggering event is a LULC event; that is the start date is the arrival of the first aquaculture ponds in an estuary. The selection of an arbitrary date would lead to an opposite set of results and conclusions.

To demonstrate the importance of longer-term analysis, in Cojimies a traditional analysis starting around the year 2000 would depict a thriving mangrove forest environment with mangroves increasing by approximately 50% over the analysis period. The longer-term reality revealed by this remote sensing analysis is that even with this substantial recovery the mangrove forests of Cojimies Estuary are still only at approximately 15% of their 1970s level and any recovery is nascent at best (Fig. 5.05). That is, an 85% decrease in mangrove forest cover would be replaced by a 50% increase in mangrove forest using an arbitrary start date or a start date based on instrument or data availability. This pattern of being misled by starting the remote sensing analysis based on instrument and data availability would be replicated in seven of the eight estuaries with only Cayapas-Mataje opposing this pattern. For example, using a Chone Estuary analysis start date around the year 2000 would be even more misleading than doing the same with Cojimies Estuary. In 1998 Chone Estuary had 1060 ha of mangrove forest, and this had more than doubled to 2165 ha of mangrove forest by 2014 (Fig. 5.07), and the estuary may be viewed as undergoing a massive mangrove expansion. The reality is that in the initial survey of 1968 Chone Estuary had approximately twice the level of the mangrove forest it has now despite this recent recovery in mangrove forest area (Fig. 5.07).

The spatiotemporal resolution used in this remote sensing analysis additionally allows for the detection of patterns not possible without such granular data. For example, instruments with either coarser spatial or temporal resolutions would not have been able to detect the LULC-change replacement type as poorer spatial resolutions would have made delineation of small shrimp farms impossible and coarser temporal resolutions would have made monitoring the LULC transitions across time highly uncertain. Additionally, the high spatiotemporal resolution of this analysis helped confirm the issue of property rights enforcement in Chone Estuary, the discovery of new fringe mangrove forests in Chone Estuary, the success of replanting projects in Chone Estuary, the creation of illegal shrimp ponds in Cayapas-Mataje Estuary, and the later removal of these shrimp farms in Cayapas-Mataje Estuary. Future sub-meter surveys, potentially by drone systems may reveal new spatiotemporal patterns that at this point remain unknown and may allow for ever increasing the human-dimension of such remote sensing analyses.

6.3 A Unique Opportunity: Global Actions, Local Responses, and the Reforesting of Ecuador's Estuaries

Despite the massive mangrove forest loss reported in Chap. 5 of this manuscript, a period of stability in the mangrove forest area of Ecuador, and even some marginal gains in the mangrove forest area of Ecuador are present in these data and can be extracted. Across all estuaries, mangrove forest loss has stabilized at close to zero loss post-2000. In some regards, this is to be expected as in some of the northern estuaries few if any remaining mangrove forest remained to be removed and rapid rates of deforestation were no longer possible. This fact of mangrove forest depletion levels by 2000 being so close total removal that it became not worthwhile to deforest anymore does not though explain some of the new mangrove forest growth that is occurring in many of the estuaries in recent surveys. For example, Chone Estuary has new and vast homogeneous mangrove forests that have appeared in the last 20-years. These new mangrove forests are not due to instrument resolution changes or other methodological factors but are genuinely present in the estuary. The post-2000 recovery of mangrove forests is a real phenomenon and is not only present in Chone Estuary but also in Cojimies Estuary, Muisne Estuary, Guayas

Estuary, and to a lesser extent on Isla Puná. Also, the pristine mangrove forest in Cayapas-Mataje has remained close to stable post-2000. The reasons behind the appearance of these new mangrove forests are worth exploring.

The reasons behind the mangrove forest growth since 2000 along much of the Ecuadorian coastline is complex and likely driven by local actions, national frameworks, and global priorities. The author of this manuscript visited the substantial new mangrove forest growth in-and-around Isla Corazón in Chone Estuary on multiple occasions while preparing this manuscript. Interviews with the local fisherman and a PPGIS workshop revealed that most of the planting had been conducted entirely at the local level and for local reasons. It was reported that the fishing syndicate on the northern banks of Chone Estuary had transitioned over both the 1980s and 1990s from fishing for income and food to providing tourist excursions for income. This livelihood transition, the fisherfolk stated, was driven by the declines in wild-catch within the estuary that the community blamed on the loss of mangrove forest due to shrimp farm expansion.

The local fishing communities of San Vicente on the northern banks of Chone Estuary had applied for and gained legal rights to parts of the Chone Estuary and have planted mangroves in this area for approximately the last 15-years. The community members stated that they wanted to return to fishing activities and they believed that replanting mangrove forests were going to be the key to their livelihood rehabilitation. The maps resulting in this manuscript reveal that their mangrove replanting effort has resulted in approximately 500 ha of additional mangrove forests being detectable by remote sensing in their planting region since the late 1990s. The mangrove replanting project, although locally driven and operated is intermittently supported by at least two US-based NGOs that provide student labor to assist in the replanting. Additionally, a representative from the government in Quito visits the community regularly to ascertain their needs, and the local office of the MAE are supportive of their mangrove replanting project.

Local mangrove planting efforts and post-2000 resistance to shrimp farm expansion is not unique to Chone Estuary. In Muisne a local NGO group was actively replanting mangroves in the estuary and was an early protector of mangrove forests in the estuary by establishing a small private and protected area of mangrove forest. In Cayapas-Mataje the concheras are active planting mangrove and resisting shrimp farm expansion through local activism and political pressure it what appears, according to the remotely sensed data presented, a successful effort. In El Oro Province local cockle collectors have claimed ancestral rights to the mangrove forests and undertaken local scale mangrove replanting tasks. It appears that a magnitude of local actors is beginning the process of reversing a decades-long pattern of mangrove forest depletion across Ecuador. It is worth noting that aside from some limited mangrove regrowth in abandoned shrimp farms in Cayapas-Mataje, the regrowth of mangrove forest across all estuaries appears to be almost wholly in areas that never had mangrove forest in the initial survey or ever converted from mangrove forest to shrimp farm during any period. That is, shrimp farms rarely appear to revert to mangrove forest, and any new mangrove forest is almost always in other areas of the estuaries and not in land that was at any time designated as a shrimp farm. The remote sensing analysis results presented dictate that due to the limited amount of estuarine space, such mangrove-forest replanting projects cannot continue indefinitely as-is and at some point for mangrove forest rehabilitation to continue shrimp farms will have to be converted back their initial mangrove forest LULC.

Although the actors are local, much of the framework for the mangrove forest recovery likely resides at the national level. During the analysis period, Ecuador enshrined in its constitution the right of all native ecosystems to survive and flourish. Also during the analysis period, the national government has drastically expanded the protected areas program adding protected areas in almost all the estuaries presented. In addition to an increase in protected area designations, national government-sponsored programs such as the Socio Bosque program and the derivative Socio Manglar program appear to be providing measurable expansion of mangrove forest or at least protection of the remaining resource. From a period of Ecuadorian government support of shrimp farming pre-2000, the estuaries of Ecuador seem to have entered a period of government support of mangrove forests since this date.

It is not just community action at the local level or government policy at the national level that has likely resulted in the post-2000 transition from shrimp aquaculture expansion to mangrove forest expansion along the coast of Ecuador that is presented in Chap. 5. It is also likely that the change in global funding priorities driven by climate change concern is playing a role in the conservation and expansion of mangrove forests. As the academic community has recently begun to realize and value mangrove forests for their role in climate change mitigation the funding agencies appear to have responded. The climate role of mangrove forests is becoming more apparent and is briefly reviewed in Sect. 1.8. The broader context of international goals related primarily to climate change is likely a driving factor in the national response to deforestation and reforestation and the creation of programs in Ecuador such as Socio Bosque and Socio Manglar that are both externally funded. These federal policies likely support local actions and actors involved in mangrove forest conservation or at least allow for a legal framework to tackle mangrove loss in which local actors exist daily.

Additionally, at the global scale, international aid agencies and other global funding mechanisms appear to have undergone a transition from supporting the blue economy in the 1980s and 1990s, focused on expanding the level of aquaculture; to supporting blue carbon in the 2000s and forward to present. Such a transition in global funding priorities promotes mangroves forests due to their essential role in mitigating atmospheric carbon, whereas shrimp farms play no important carbon role. Section 2.1.1 reviews just a fraction of the international funding that likely went to support shrimp aquaculture adoption and expansion in Ecuador in the 1980s and 1990s but now these same funding agencies appear to be moving their funding away from shrimp aquaculture and towards mangrove forest conservation. This implication of the migration from international financing of the blue economy to international financing for blue carbon, as it pertains to mangrove forests and beyond, is a frontier of coastal research. Although mangrove recovery in Ecuador is relatively recent and starting from a much-reduced mangrove forest level, it does appear that local actors, national policies, and global priorities based around climate change mitigation provide a unique spatial and temporal opportunity to replenish a lost land cover within Ecuador's estuaries. This remote sensing analysis captures this nascent recovery of mangrove forest, and it is likely the first time this trend appears in the mangrove forest research literature.

6.4 The Unique Land Cover and Land Use Change Story of Cayapas-Mataje Estuary

The lack of substantial mangrove forest loss or shrimp farm expansion in Cayapas-Mataje Estuary is likely due to a number of interacting reasons. Firstly, Cayapas-Mataje was relatively isolated from the rest of Ecuador during the majority of the analysis period. It was difficult and expensive to move the iced shrimp from Cayapas-Mataje to the port of Guayaquil until a paved road connected this region to the container port. Additionally, without reliable road transportation, it would have been difficult to import the heavy machinery required to build and operate the shrimp farms or to transport the feed and other supplies needed to operate a shrimp farm. Compared to the other estuaries analyzed, Cayapas-Mataje is the furthest of the study estuaries from the only port capable of exporting shrimp in Ecuador.

The isolation of Cayapas-Mataje may have played an additional factor in the preservation of the mangrove resource; this analysis reveals that in Ecuador's northern estuaries, shrimp farming likely began in Chone Estuary between 1968 and 1977 and gradually moved north into Cojimies Estuary and Muisne Estuary over the subsequent decades of the 1980s and 1990s. Cayapas-Mataje is not only the most distant of the northern estuaries from Chone Estuary; it was also not possible to travel by road along the coast to Cayapas-Estuary from Guayaquil or the Chone Estuary until around 2000. Until this time, all road journeys had to enter the highlands and return to the coast to reach Cayapas-Mataje over a day or two. Indeed, an intermittent and slow rail journey was the primary way to get to Cayapas-Mataje until the 2000s. When I ask residents of Cayapas-Mataje Estuary how they managed to protect their mangrove resource a reoccurring answer is that many artisanal fisherfolks moved here from around Chone, Cojimies, and particularly from Muisne and these migrants had been driven out of these estuaries further south due to a loss of their livelihoods resulting from shrimp farm expansion. The migrants knew the damage shrimp farms could cause to local fisherman's livelihoods from their experiences further south and actively resisted shrimp farming once it arrived in Cayapas-Matake. The fisherfolk migrants were not only active themselves in resisting shrimp aquaculture expansion in Cayapas-Mataje but also informed the local cocklers and fisherfolk of the impending loss of livelihoods if shrimp farming was allowed to displace the mangrove resource.

The second reason for the lack of shrimp farming in Caypas-Estuary may be due to the unique protected status of almost all of Cayapas-Mataje Estuary under the REMACAN designation. Although six of the eight study-sites have some form of protected status, Cayapas-Mataje is an original RAMSAR site, and the government of Ecuador is the owner as well as the protector of the critically important wetland site and it is the only estuary in Ecuador that had longer-term protected status. Within the hierarchy of differing protected mangrove sites in Ecuador, Cayapas-Mataje has the highest level of protected status and was the only study site with this status. Cayapas-Mataje has been managed for conservation for almost the entire analysis period with a management system that explicitly targets the protection of the mangrove forest resource as its primary goal. As opposed to REMACAN, other Ecuadorian estuaries have much less stringent management goals, and protected areas that are more recent in implementation, and the newer protected areas do not always exist purely to conserve the mangrove forest resource.

As described in Chap. 3, external financing likely played a crucial role in the initial creation and expansion of shrimp farming in Ecuador and a lack of international and international investment in the Cayapas-Mataje region is likely the third reason for a different LULC-pattern occurring in this estuary. During much of the analysis period, the Colombian border region of coastal Ecuador was considered a politically unstable place and aid agencies developments, and FDI likely avoided the region. Due to FARC and other group activity, the Cayapas-Mataje region was deemed to be dangerous, and investments in the region would be considered high risk. Indeed, 2006 and 2007 travel to Cayapas-Mataje was not advised by numerous foreign governments including the US State Department as FARC activity occurred in the region and illegal activities included drug smuggling and kidnapping of foreigners was commonplace.

The final reason for the differing LULC history in the Cayapas-Mataje region, as opposed to the other estuaries analyzed, is likely cultural. Mangrove forests are at the heart of the community in Cayapas-Mataje. The Afro-Ecuadorian population has mangrove murals in their homes, they celebrate mangrove holidays including days dedicated to protecting the mangrove forests, and they are very active in conserving the mangrove resource including physically resisting mangrove clearing for aquaculture ponds. Chapter 3 touches on many of these details including how the town monument is dedicated to the products of the mangrove forest and how the residents of this region live inside the forests and not merely on the terrestrial banks of the estuary. The concheros and the fisherfolk of Cayapas-Mataje appear to treat mangrove conservation as a way of life as opposed to even just a livelihood. Ocampo-Thomason (2006) provides a far more complete review of this way of life that this remote sensing analysis can accomplish. What is worth saying, is that in many ways the cultural norms and societal structures of REMACAN when compared against the other estuaries is captured by the remote sensing analysis presented of the researcher is capable of making the connections.

6.5 Policy Recommendations

Although broad policy recommendations are beyond the realm of this manuscript, it is worth touching on a couple of policy issues that this remote sensing analysis reveals. Firstly, at the local, national, and international level aquaculture and fisheries are often considered a single activity with a singular goal. This is best represented at the global level by the United National Food and Agricultural Organization Department of Fisheries and Aquaculture that represents both fisheries and aquaculture. This study reveals that commercial aquaculture was almost the sole reason for the loss of mangrove forest in Ecuador. It is well-publicized that mangroves in Ecuador and globally are the driver of many important nearshores such as non-farmed shrimp, numerous crab species, numerous mollusks, mullet, and others as well as providing breeding grounds and juvenile habitat to numerous commercial offshore fisheries. The investment in shrimp aquaculture in Ecuador during the 1980s and 1990s may have led to an increase in Whiteleg Shrimp, but it likely led to the decline of many other commercially essential fisheries on which the locals relied for food and employment. In this regard, the investments made into shrimp aquaculture may have harmed many wild fisheries and in turn, resulted in increased hardship for the local fisherfolk and even potential declines in the commercial offshore fishing industry. A simple solution may be to sperate fisheries support from aquaculture at all scales as opposed to having them in a single framework as the actions of one may cause losses in the other. Indeed, it is worth asking the question of whether international agencies and development banks at any scale should be in the business supporting shrimp aquaculture or any aquaculture and the expense of mangrove forests and wild fisheries.

The final policy point is that this study reveals that shrimp farms rarely if ever return to mangrove forest cover. This reasons for this are likely that shrimp farms would need to be purchased for mangrove reforestation whereas shallow estuarine water, mud-banks, and salt-pans do not. Additionally, even after abandonment former shrimp farms do naturally return to mangrove forest for a host of reasons including the change in the properties of water flow in the estuary, the high levees around the ponds that would need to be removed, and the potential for other invasive vegetation to be present. Potential remedies could include paying for rehabilitation of abandoned ponds, offerings incentives for conversion of productive farms, or agreeing of transfer of terrestrial lands for ponds in return for aquatic lands. Despite numerous government decrees that attempt to enforce conversion from shrimp ponds to mangrove forests and laws that shrimp farmers must convert a percentage of their ponds back to mangrove forest; this study reveals that this conversion is not occurring and further incentive policy actions may be required.

6.6 The Next Fifty Years

The next 50-years will undoubtedly bring a new set of LULC changes to Ecuador's estuaries. Any future LULC-transitions will continue a long tradition of boom-and-bust for the mangrove forest of this region. From the colonial period of mangrove forest timber extraction for ship construction to mid- twentieth century mangrove forest clearing driven by the tannin industry, to the post-1970 period of mangrove clearing driven by shrimp aquaculture the mangrove forests of Ecuador have likely always been a dynamic and changing environment. The difference between the ship construction and tannin clearings is that the mangrove forests are not merely cleared for shrimp ponds but entirely replaced by a new commodity-driven LULC that does not allow mangrove forest regrowth once the commodity is extracted. Another critical difference between the earlier mangrove forest clearings and the more recent mangrove forest clearings is the ability of researchers and policy-makers to track and respond to the LULC-changes in almost near time via rapid forest loss reporting via remote sensing.

Future remote sensing surveys, likely until at least 2030 due to mangrove growth rates, will be required to ascertain if the period of post-2000 mangrove recovery in Ecuador is indeed a trend of true longer-term mangrove forest recovery or a brief period of respite in a longer-term pattern of mangrove forest declines. Such future surveys will have access to new high-resolution government instruments such as Sentinel-2, even-higher resolution privately-owned cluster satellite systems such as those offered by Planet, and even higher resolution researcher-owned systems such as those now used by researchers mapping mangroves from drones. Undoubtedly new and yet unknown EOS systems and more advanced processing methodologies will also be developed. Each new system will need to be integrated into the long-term analysis presented to allow for comparison to the earlier datasets, and undoubtedly the higher spatiotemporal resolutions of these systems will reveal unknown LULC patterns in the estuaries. The future of Ecuador's estuaries and mangroves forest is difficult to predict but what is certain is that the LULC changes will be monitored and reported on, and the period of time-consuming and costly recreation of prior landscapes is closing. The future challenge will be for coastal remote sensing researchers to combine and integrate their findings and research with other disciplines to fully capture the human-dynamics of remotely sensed LULC-change.

References

Boyd CE, Clay JW (1998) Shrimp aquaculture and the environment. Sci Am 278(June):58–65

CGMFC-21, Global Mangrove Levels by Country 2000–2014 (2015) Harvard dataverse. https://doi.org/10.7910/DVN/HS5OXF

Giri C, Ochieng E, Tieszen LL, Zhu Z, Singh A, Loveland T, Masek J, Duke N (2011) Status and distribution of mangrove forests of the world using earth observation satellite data. Glob Ecol Biogeogr 20(1):154–159. https://doi.org/10.1111/j.1466-8238.2010.00584.x

Hamilton SE (2015) No evidence that shrimp aquaculture is responsible for minimal mangrove deforestation. Bioscience 65(5):457–457. https://doi.org/10.1093/biosci/biv030

Hamilton SE, Casey D (2016) Creation of a high spatio-temporal resolution global database of continuous mangrove forest cover for the 21st century (CGMFC-21). Glob Ecol Biogeogr 25(6):729–738

Hansen MC, Potapov PV, Moore R, Hancher M, Turubanova SA, Tyukavina A, Thau D, Stehman SV, Goetz SJ, Loveland TR, Kommareddy A, Egorov A, Chini L, Justice CO, Townshend JRG (2013) High-resolution global maps of 21st-century Forest cover change. Science 342(6160):850–853. https://doi.org/10.1126/science.1244693

Ocampo-Thomason P (2006) Mangroves, people and cockles: impacts of the shrimp-farming industry on mangrove communities in Esmeraldas Province, Ecuador. In: Hoanh CT, Tuong TP, Gowing JW, Hardy B (eds) Environment and livelihoods in tropical coastal zones: managing agriculture-fishery-aquaculture conflicts, Comprehensive assessment of water management in agriculture series, vol 2. Oxford University Press, London, pp 140–153

Printed by Printforce, the Netherlands